# LIFE HISTORY

OF

# OUR PLANET.

WYOMING IN THE LATER EOCENE. (See page 249.)

# LIFE-HISTORY

OF

# OUR PLANET.


By WILLIAM D. GUNNING.

ILLUSTRATED BY MARY GUNNING.


"They say
The solid earth whereon we tread,
In tracts of fluent heat began,
And grew to seeming random forms,
The seeming prey of cyclic storms,
Till at the last arose the man;

Who throve and branched from clime to clime,
The herald of a higher race,
And of himself in higher place,
If so he type this work of time
Within himself."

CHICAGO:
W. B. KEEN, COOKE & CO.
1876.

# PREFACE.

WITH much of that which is called "Popular Science" the writer of these pages has no sympathy. To read from pages addressed to the memory only, that the sun is so many miles from the earth, is so many miles in diameter, and is enveloped in "whirlwinds of tempestuous fire" so fierce that if they could blow upon the earth they would reduce it in a few seconds to a mist of atoms, may help the mind to astronomical facts and set it aglow with wonder, but it does not teach it *astronomy*. To read in similar manner that rocks were formed so, or so; that oceans came to be in this wise or that wise; that continents rose through the play of these forces, or those; that life began in one form or in another, and to read how it unfolded and branched from order to order, may be entertaining, but it may have little to do with *geology*.

Facts do not enlarge the mind unless they are fertilized by principles.

Our aim in the preparation of this volume has been to conduct the reader through methods to results. The leading types of life which have possessed the earth from age to age he will find described and delineated. He will find the more significant types reconstructed, part by part, with so little of the phraseology of comparative anatomy, that his mind, it is hoped, will traverse the methods and make them its own.

To all who have labored in the quarry, the museum, the laboratory, or the library, and who are known to him by their works, the writer acknowledges his indebtedness.

W. D. Gunning.

Waltham, Mass., Oct., 1876.

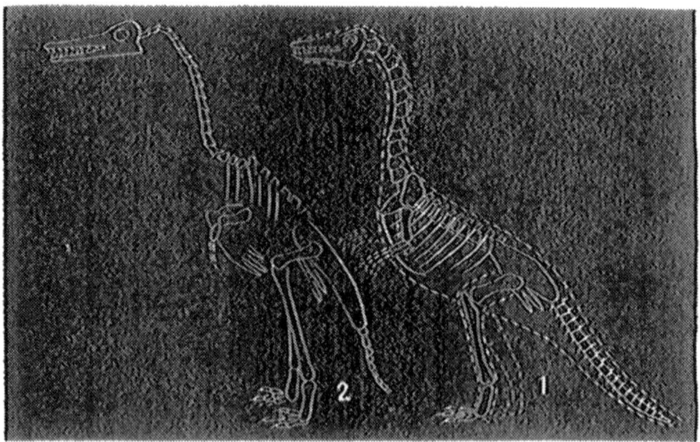


On the Way to a Bird.—1, A Dinosaur of the Jurassic Period—a Bird except in teeth, fore limbs and tail. 2, A Bird which lived in Kansas during the Cretaceous Period—a Bird in all except *teeth*, which lingered as vestiges of the Reptile.

# ILLUSTRATIONS.

# TABLE OF CONTENTS.

## CHAPTER I.



## CHAPTER II.



## CHAPTER III.

## CHAPTER IV.



## CHAPTER V.



## CHAPTER VI.

## CHAPTER VII.



## CHAPTER VIII.

## CHAPTER IX.

# OUR PLANET:

## ITS LIFE HISTORY.

---

### CHAPTER I.

Activities through which Nature came to Rest in the Rocks — History of Rocks and Metals in Molecules — History of Rocks in Masses — A Molten Globe — A Globe Cooling and Shrinking — A solid Core and a solid Crust — Reactions between the Molten Zone and the Crust — Mountains, Earthquakes, Volcanoes — Volcanoes Incident to a Globe Advancing in Age — Earthquakes Incident to a Globe in its Youth, and Decline as the Globe advances in Age — Solid Nucleus meets Solid Crust and Nature comes to Rest in Rock-masses — A World in its Old Age and Death.

IN ROCKS NATURE IS AT REST.

Strip from the earth all animals, all plants, and whatever lives and is neither animal nor plant, and you will have a mineral world. Break from its crust, a fragment and make it an object of study; you will find no interdependence of parts. You cannot see the whole animal in a limb, nor the whole plant in a leaf; but you can see all the iron of a mine in a bit of its ore, all the granite of a ledge in a detached fragment, all the ocean in a water drop. But while you find no interdependence of parts, you will find a union of elements. Oxygen is an element which turns like the point of a dagger toward everything that lives, and its incessant attack on leaf and blood occasions all movement in the organic world. But rocks, being already saturated with oxygen, are free from attack. In them the elements have satisfied their affinities, and are therefore at rest.

But the rest to which nature has attained in the rocks implies antecedent activities. The cinders of a burned city tell you of oxygen whose likings have been gratified at the cost of a conflagration.

Rocks are the Cinders of a Burned World.

These cinders tell you of oxygen (enough to form nearly half the earth's weight) whose affinities have been gratified at the cost of prodigious energies.

Rocks are the Product of Spent Forces.

Nature, as the name implies, is a continuous on-flow. She is not fixed but fluent. As fixed ice to the fluent sea, so are rocks to nature. They are nature at rest.

Fig. 1.

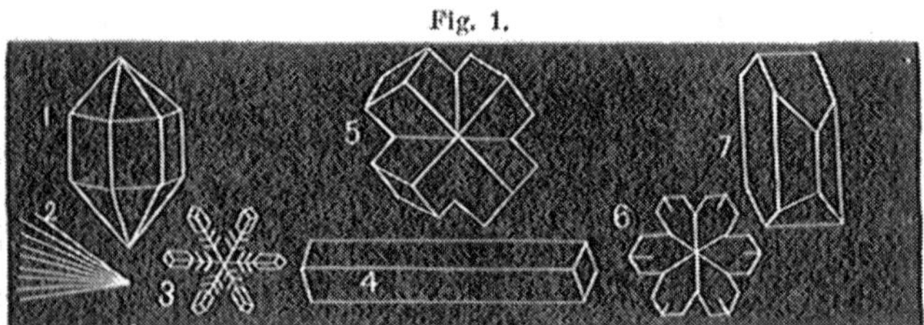


Rocks are Historic Records.

Each element of a rock mass has tried to realize an ideal of form. When water passes into the rigidity of ice, its molecules compose themselves to rest in the form of beautiful six-rayed stars. When the molecules of silver are freed from nitric acid they

Fig. 1. 1. A crystal of quartz. 2. A bundle of hornblende crystals. 3. A crystal of water. 4. Of actinolite. 5. Of staurolite. 6. Of water. 7. Of feldspar.

march in most wonderful order, each to its allotted place of rest in a silver tree. Latent in every molecule of water and in every fluent molecule of metal or mineral lies this marvelous building power. Active once in every molecule of every crystal was this building power.

Silicon is an element which does not exist alone, having everywhere satisfied its affinities for oxygen and become *quartz*, which, in some form, constitutes about one-fourth of the earth's crust. The molecules have built themselves into prisms, terminated by pyramids. The facets are always six, and however the form of the crystal may vary — and it does vary from almost perfect symmetry to sheets as thin as the blade of a knife — the angles are always the same and the number of facets always the same.

Silica, which is silicon combined with oxygen, has united with alumina and an alkali and formed *feldspar*, which, after quartz, is the most common of minerals. The molecules have marshaled themselves to rest, forming crystals, commonly, of ten facets.

Silica has combined with alumina and potash or magnesia and formed *mica*, whose molecules have united to make thin, transparent, brilliant, crystalline leaves.

Silica has combined with magnesia, oxide of iron, and lime, and the result is *hornblende*, a mineral whose molecules have formed dark, oblong prisms, or tufts of green needles, called actinolite; or white, called tremolite; or bundles of white fibers, called asbestus.

These are the principal minerals which compose the earth's crust. Silica, alumina and potash or soda, magnesia, oxide of iron, and lime, when held in solution by water at temperatures above 700° F., crystallize — the silica into quartz; silica, alumina and potash or soda, into feldspar; silica, potash and magnesia, into mica; and silica, magnesia, lime and oxide of iron, into hornblende. We infer that rocks which contain these crystals, or any of them, were formed through the agency of water, and at a temperature which may not have been higher than seven hundred degrees Fahrenheit.

Rocks are either crystalline or fragmental. Quartz, feldspar, mica and hornblende, when crystallized separately in grains fine or coarse, and united into a compact mass, form the crystalline series of which granite is the type.

Pebbles of quartz, grains of quartz, and grains of feldspar and mica, when cemented, pebble to pebble, and grain to grain, form conglomerate, sandstone and slate — a series of fragmental rocks. Fragmental rocks cannot be the older as they are only cemented fragments torn from other rocks. Granite cannot be the *oldest*, as its constituents assume their crystalline form at low ranges of temperature, and as it often contains talc and chlorite which are *hydrous* minerals and show that water was present in the granitic paste.

The oldest rocks would tell us of intense heat, for all lines of search lead back toward a beginning when our globe was incandescent. The first step nature took in rock-making was the formation of the stable oxides, that is, the burning of silicon, aluminum and the like into their cinders, *silica*, *alumina*, and the like. But the crust which first hardened over a molten, incandescent globe nowhere appears. The oldest rocks are everywhere buried under their own ruins.

When the world was incandescent where were the oceans? Not *on* the earth but in the air *above* it. No matter has been added — save in the fall of meteorites — and none has been taken away. The waters were here from "the beginning," here in their elemental gases. As silica is the oxide of silicon and implies a time when its elements were free, so water is the oxide of hydrogen and implies a time when its elements had not combined. Water is a cinder from the burning of gas with gas, as silica is a cinder from the burning of a gas with a solid.

When the earth and its atmosphere had cooled to the dew-point, condensation occurred. *Oxidation* must have occurred before, and the oceans must have hung as a pavilion of clouds and a thick veil of mist over the earth before they fell in rain upon its surface. They did not fall alone. The primeval atmosphere was loaded with other vapors than those of water. The vapor of every element which is volatile at a high heat would have filled the sky. Chlorine and sulphur and carbon, in the form of acid gases mixed with watery vapor, would have formed a dense atmosphere. These acids would have come down in the rain, entered into combination with certain bases and loaded the sea as they had loaded the air.

We are not left to conjecture. We have portions of the primitive sea imprisoned in pores of the rock formed on its bottom. This fossil sea-water, as Dr. Hunt has shown, is far richer than our seas in the salts of lime and magnesia, but not in common salt. Most of the chlorine in the present seas has entered into combination with sodium and formed common salt, but half the chlorine in the primeval sea was combined with lime and magnesia — a fact we shall find of great significance in the after-history of the globe.

Gold and silver and copper and lead are held in solution by the sea. Copper is found in the blood of marine mollusks, and copper, silver, and lead are found in the ashes of marine weeds, while a German chemist has announced the discovery of gold in sea-water. Sonstadt's experiments have led him to estimate one grain of gold to each tun of ocean, or one dollar to each twenty-five tuns, and if this estimate is correct the oceans hold eleven million two hundred thousand times as much gold as all the nations of men have dug from the earth!

These metals were held in solution by the primeval sea in larger stores, we have reason to believe, than they are contained in the seas of our own time. What has become of the excess?

History is recorded in metals as well as in rocks and oceans. If the rest to which nature came in crystalline granite implies vast antecedent activities, the metal she has brought to rest in strata and veins and fissures of the earth is the inscribed record of activities equally great.

What quartz is among rocks, that iron is among metals, the most abundant and the most universally diffused. It forms, possibly, the core of the globe to within five hundred miles of the surface. It appears, *certainly*, as a large constituent in the deepest known rocks, and its ores vein the oldest rocks which come to the surface. It is as old as its containing rock.

What is its history? Symbols will help the mind to traverse

THE HISTORY OF AN ATOM OF IRON.

It is the province of science to make the mind see where the eye cannot. Our eyes cannot see an atom, but *we* must see it. We symbolize the iron atom by Fig. 2 at 1. We cannot question it as to its essence or origin. We do not know that it ever had an origin, and we must take it as the mathematician takes his numbers. The first step we find nature taking with that atom is *to burn it.* The atom combines with one and a half atoms of oxygen and is then a *molecule.* Our molecule is insoluble in water. If it were in the sea it would find its way to the bottom and rest there in the mud. If this oxidation occurred when the world was young and there was yet "no sea," the molecules would lie as cinders on the earth. They would be diffused through the rocks forming at the time of their oxidation. They *are* diffused through almost all the rocks of the earth's crust. And as gravel and clay and sand and soil are merely abraded rocks, they are diffused through almost all the top-dressing of the rocky crust. As they are insoluble in water, the rain which sinks through tho soil has no power to pick them out and carry them away. But if the rain falls on a soil overgrown with vegetation the result may be different. Moss, ferns, grass, leaves — all forms of vegetation — while rotting, yield portions of their carbon to the water which falls on them. Freighted with this carbon the water sinks into the soil, and every molecule of iron it finds on the way it picks out and carries with it. For the molecule gives up one portion of its oxygen to the carbon in the water and forms with it carbonic acid. What remains now is a *protoxide* — one atom of iron to one of oxygen — and this is soluble in water containing acid. The water can move the molecule as soon as this interchange has occurred.

Fig. 2.

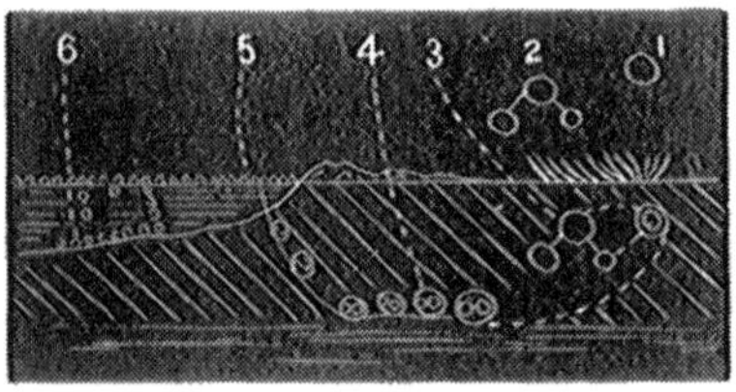

If we look now on a stagnant ditch in a meadow, or a stagnant pool on a wayside where vegetation is rank and rotting, we may see it mantled with an iridescent scum. This scum is the iron picked from the soil and carried hither by the acidulated water. The molecules which lay against the air took from it each another portion of oxygen, and, becoming again insoluble, separated from the water and formed this thin, shining film. After a time the film sinks to the bottom where it is followed by other films which appear now as a reddish ochre. If the water comes up in a bog the iron-molecules it brought are precipitated as bog-ore.

We have found the atom oxidizing and becoming a

Fig. 2. History in Symbols of an Atom of Iron.—1. An atom. 2. The atom having combined with a portion and a half of oxygen appears in a molecule of peroxide. 3. The molecule diffused through rocks or soil, and yielding a portion of its oxygen to the carbon which a water-drop, represented to the right, has carried down into the soil from decaying vegetation above. 4. The iron molecule, reduced now to a protoxide, is taken from the soil by acidulated water and borne to 5. A bog or spring or pond. 6. The molecules of protoxide which lie against the air, absorb from it another portion of oxygen and becoming *peroxide*, separate from the water and unite to form a thin iridescent film. The film, growing thick, falls to the bottom and appears as iron-ochre or bog ore — the first stage of all iron ores.

molecule. We have seen molecules *diffused* through rocks and soil. We have seen them giving up to the carbon of decaying vegetation, brought to them by water, a portion of their oxygen and becoming soluble. We have seen them taken up then by the water and carried to ditches or ponds or bogs and there *aggregated* as iron-ochre or bog-ore. By other agencies this ochre or bog-ore is changed into any of the forms of iron found in the earth's crust.

The shining film on the pool will give us a vision of other times. For as nature works to-day in gathering up and aggregating her molecules, so she has always wrought. A little bed of iron-ochre on the bottom of a ditch tells you that so much vegetation had to grow and die and rot and yield its carbon to percolating waters, before so many molecules of iron could be gathered from the soil and brought together on the face of a ditch. A mountain of iron-ore — as that in Missouri — tells you that so much vegetation, the same in proportion as the mountain to the ochre, had to grow and die and rot on the face of a primeval world, and yield its carbon to be burned by iron-molecules and thus to acidulate the percolating waters, before so many particles of iron could be gathered from the ground and aggregated in a bog or under stagnant waters.

The iridescent film we see so often we do not see at all, may give wing to our imagination and show us a vision of other worlds as well as other times. A law here is a law of the universe. Our minds are so made that when we see the results of a process going on before our eyes, and then see the same results brought to us, even from interplanetary spaces, we must assert a like process for a like result. On the twelfth of February, 1865, ten o'clock at night, a hundred and fifty miles above Iowa, there appeared a meteor two thousand feet long and four hundred feet in diameter. It shone like a sun and illuminated the entire State of Iowa and portions of Missouri, Illinois, Wisconsin and Minnesota. Men and animals were smitten with dread. It shot athwart the sky at a speed of twenty-one miles a second, and exploded at a height of ten miles with detonations which shook the houses far and near. The meteor contained by weight seven per cent. of metallic iron. Meteoric stones, then, we are constrained to say, are waifs from another world, perchance a dismantled world, in which vegetation was performing the same functions in aggregating molecules of iron as it performs here.*

Less abundant than iron is *lead*, but not less interesting as an episode in the earth's history is

### THE HISTORY OF AN ATOM OF LEAD.

Lead is held in solution by the seas, and was held in larger per cent. by the primeval seas. Where is the excess? How has it been abstracted from the sea and laid to rest in the limestones?

Prof. Whitney prepared the way for an important

* "The presence of graphites, of native iron and of sulphurets in most ærolites, tells us in unmistakable language that these bodies come from a region where vegetable life has performed a part not unlike that which it plays on our globe, and even leads us to hope for the discovery in them of organic forms which may give us some notion of life in other worlds than our own." —*T. Sterry Hunt. The Chemistry of the Earth. Rep. Smith. Inst.*, 1862.

reform in geology when, in a government report on the lead region of Wisconsin and Illinois, he called attention to the fact that metals had been introduced into the rocks of the Northwest at the very time when life was introduced on a grand scale into the oceans of the globe. From the *oceans* came the lead — not from an imaginary ocean of lava at the earth's core. Suppose we had a tank, filled with sea-water. And suppose we were to introduce a solution of lead until the water is saturated with it. The lead must be in a form known as *sulphate*, a compound of the metal and sulphuric acid. Symbols will help us, and we write the chemical symbol of this compound, $Pb.\ S_4$., which we are to read, lead combined with sulphur combined with four portions of oxygen.

Lead in this form the sea can hold in solution, and in this form we are to suppose all the lead was held while it was yet fluent. Our *tank-sea*, heavy with sulphate of lead is an epitome of the primeval sea. We may cover it over to prevent evaporation, and then wait and watch for the appearance of lead. But no lead is apparent. Not a spangle appears on the walls of the tank. It is evident that the lead cannot help itself out of the sea. Calling to mind the fact that lead was introduced into the rocks when life was introduced into the sea, we stock our tank with sea-weeds and low forms of animal life, like polyps and and medusæ. We watch and wait again, but no lead is precipitated. It is evident that *life* cannot help the lead out of the sea. We wait till death has occurred — the death of a plant, or of some animal, as a polyp or jelly fish whose body is composed chiefly of carbon. The glass would now reveal minute cubes of lead spangling the floor of the tank. As deaths are multiplied the crystals grow larger. It is evident that in some way death helps the lead out of the sea into the rocks on its bottom. How? As the molecule of iron gave up a portion of its oxygen to the carbon of the dead plant and became *soluble*, so the molecules of lead give up their oxygen to the carbon of the dead, and become *insoluble*. We change the symbol now and read Pb. S., which means that a lead atom has combined with a sulphur atom. This form of lead we call plumbic sulphide, or galena. Being insoluble it sinks and crystalizes in cubes on the floor of the sea. Currents may sweep it into fissures in the rock, and after-changes wrought on the containing rock may modify its form or chemical structure.

Fig. 3.

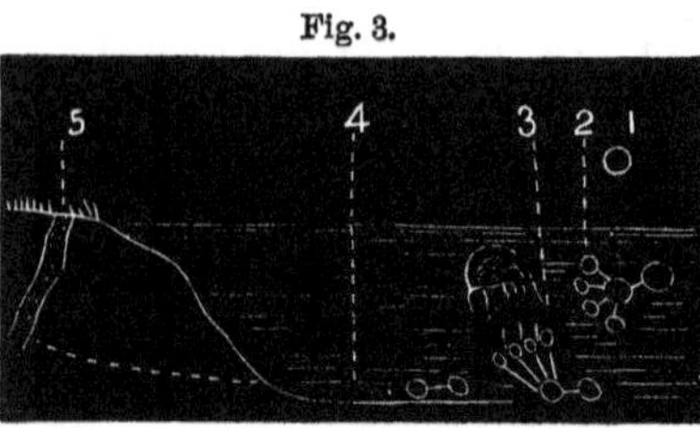


Very different from lead in its properties is *copper*, but not very different from the history of lead is

THE HISTORY OF AN ATOM OF COPPER.

Copper, like lead, is both fluent and fixed. Held in solution by the seas it circulates with their waters;

Fig. 3. HISTORY IN SYMBOLS OF AN ATOM OF LEAD.—1. An atom. 2. The atom combined with sulphuric acid and appearing in a molecule of plumbic sulphate held in solution by sea-water. 3. A medusa which, at death, takes the four portions of oxygen from the molecule and leaves a molecule composed of one portion of lead and one of sulphur. 4. This molecule, called now *plumbic sulphide*, being insoluble is precipitated. Such molecules are diffused through the sediments of a sea-bottom and through rocks which were sediments. 5. The molecules segregated from the strata and brought together in a vein.

locked up in the crystalline rocks, it has been withdrawn from circulation and laid at rest.

The copper-atom combined with sulphuric acid and formed a molecule which we express by the symbol Cu. $S_4$. It reads: Copper combined with sulphur which had combined with four parts of oxygen, and is called *cupric sulphate.* This is the most common salt of the metal and is soluble in sea-water. In this form the present seas hold copper and in this form the primeval seas held it in larger store. The copper stored in the rocks is a measure of the excess. Our imaginary tank-sea may serve us again. We must suppose it to hold in solution a large per cent. of this copper-salt. No metal will be precipitated until the sea is tenanted. We introduce sea-weeds, and mollusks, and polyps. No metal is apparent until death has occurred, and then little crystals or spiculæ of *sulphide* begin to accumulate on the floor. The sulphate yielded its oxygen to the carbon and became insoluble cupric sulphide. So far the history of a copper-molecule is the same as that of a molecule of lead. But another process is

Fig 4.

Fig. 4. Doris lacina.

going on in the sea. Living mollusks have greater chemical powers than dead plants or polyps. Strange that a sea slug like that depicted on the preceding page, beautiful in its horn-like antennæ, and lung like crimpled fronds of a fern — beautiful but low in structure and vitality — strange that its sluggish life should break up the molecule and take the copper atom from the strong embrace of the acid and make it a part of its own blood. But this power is given to every marine mollusk. It breaks up the molecules of sulphate and takes the naked atoms of copper into its blood. If our tank-sea were tenanted by mollusks they would, in time, abstract all its copper and lay it down in insoluble grains on the floor. And if it were the great sea itself, and had been tenanted by mollusks through countless ages, diffused through all its depth of sediment would be the copper which had pulsed through the blood of all its mollusks. And if it were the seas that had been, and its sediments were transformed into rocks, diffused through these rocks would be all the cupric sulphide and all the native copper precipitated by dead and decaying

Fig. 5.

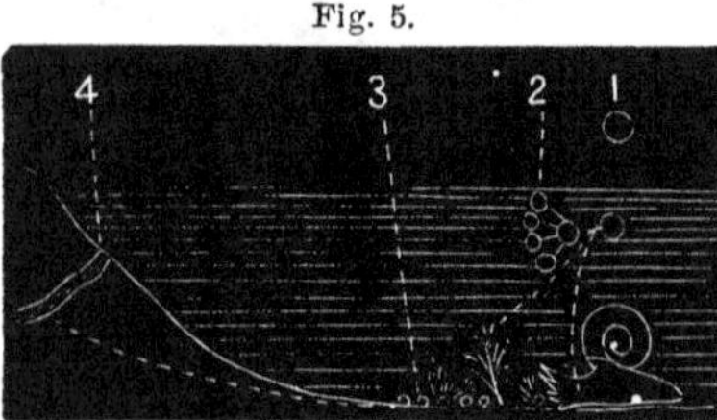


Fig. 5. History in Symbols of an Atom of Copper.—1. The atom. 2. The atom united to an atom of sulphur which itself has united with four atoms of oxygen. Such a molecule of cupric sulphate is held in solution by sea-water. Mollusks and sea-weeds, represented at 3, break up the molecules and take the copper, the mollusk into its blood and the weed into its tissues. Copper, taken thus from the seas is diffused through sediments and sedimentary rocks. 4. Copper segregated from the strata and *aggregated* in a vein.

organisms, or assimilated into the blood by living organisms. Diffused once as a mist of atoms through a nebula, diffused again as molecules through an ocean, the copper would be diffused now as invisible particles through the rocks. The history does not end here. The diffused particles must be picked out of the rocks and brought together into masses. We shall understand this process of aggregation after sketching

### THE HISTORY OF SILVER AND GOLD.

*Pure* silver is no more currency in nature than pure lead or copper. Before the sea can hold it its atoms must enter into combination with grosser elements. Sulphur which gives currency to so many metals makes silver current. In the form of a sulphate its molecules are fluent and circulate through the seas.

On the shallow floor of tropical seas are gardens of coral-making polyps. Bischoff has lately taken silver from the bodies of reef-building polyps, proving that these animal-flowers have the same power to break up the molecule of silver as mollusks to dissolve the molecules of copper.

On the shallow floor of all seas, and on their tide-washed shores, and often "banking mid-ocean," are gardens of algæ. These humble plants are gifted with the same powers as the humble polyps. Bending to and fro with the restless sea that bathes them, they break up the restless molecules of silver, and take the metallic atoms into their own fibre. Polyp and sea-weed withdraw silver from circulation and retire it. The sediments of all seas in which algæ have grown, and all polyp-made limestones should be the rich banks of deposit for the silver withdrawn from circulation during the geologic ages, by plant and polyp. And they are. But the metal is *diffused*. It must

Fig. 6.

enter again into the circulation of nature and be aggregated and stored away in vaults of rocks before man can extract it and store it in his own vaults.

Gold is held in the sea, held there soluble and oxidized, as shown by the experiments of Sonstadt, by iodine. It is precipitated and strewn in fine particles through the bottom, by the products of organic decay.

Fig. 6. Spiridea—a Sea-weed.

Gold, like copper and silver, when withdrawn from circulation is diffused, and, like them, it awaits aggregation. By what process is this achieved?

A cooling globe shrinks and cracks. Age after age the crust of our cooling earth has broken and opened in fissures long and deep. But for her powers of surgery by which nature has healed these fractures we should find the earth's surface hardly fit for habitation.

Fig. 7.

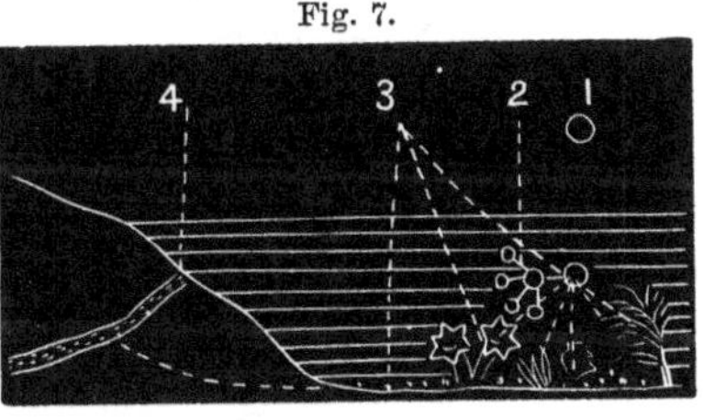


See now how wisely she heals the rents and gashes Time is forever making in her face. Water circulates through the earth's crust. It is almost a universal solvent. It takes up particles of limestone in its flow and bears them along to an open fissure where, on evaporating, it throws them down against the walls of rock. It takes up alumina and under rare conditions throws the atoms down against the walls of the fissure in crystals of sapphire. It takes up alumina and silica and fluoric acid and under conditions equally rare it drops the molecules into the open fissure in crystals of topaz. In the same way it may gem the fissure with tourmaline and garnet and emerald and amethyst and hyacinth. When, deep down, it becomes thermal water and holds in solution sulphides and alkaline car-

Fig. 7. History in Symbols of an Atom of Silver.—1. The atom. 2. The molecule held in solution by the sea. Silver is withdrawn from the circulation of nature by polyps and sea-weeds, represented at 3. Silver abstracted thus from the sea, is diffused through sediments and sedimentary rocks rich in the ruins of sea-weeds, and through rocks made by polyps. It is separated from such strata by thermal waters and brought, atom to atom, into veins represented at 4.

bonates, it dissolves from the rocks the particles of metal they contain and brings them to the open fissures. Rising in these fissures and losing heat, it loses its power to hold the minerals and metals it brought. They are thrown down and help to heal the fracture.

So veins have been filled from the earliest ages, and so they are filling now. So thermal waters in Nevada are seen to rise through fissures and throw down minerals and metallic ores along their walls. So thermal waters are known, by the metallic ores they have left, to have risen in fissures and veins that opened in the earth's crust æons ago. And so the rents and gashes in the crust have been healed—healed by a surgery analogous to that by which nature mends a fractured bone in our own frame. Wonderful this mimicry of life!—as if the globe were a great organism with rills of water coursing through it like rills of blood through a body, and picking up an element here and another there, as blood in the body, and bearing them away and massing them, iron here, and copper here, and silver here, and gold here, as blood carries lime to the bones and phosphorus to the brain. Wonderful and beneficent this mimicry of life, beginning in the infinite past and through all the æons recorded in the globe's crust healing its wounds and aggregating its metals for the use of man.

"And I doubt not through the ages one increasing purpose runs."

The analogy is closer than we have named. A world, like an organism, passes from youth into old age and death. In youth our globe was molten. From youth till now it has been a cooling globe; from now till the end it will be a cooling globe. A molten sphere, in cooling, contracts and wrinkles, and generates earthquakes and volcanoes. From the history of the parts we turn to the history of the mass.*

As the matter of a molten globe presses toward the center, solidification, as Hopkins has shown, would begin there. *Pressure* would cause solidification at

* We make no pretense to an exhaustive treatment of the origin of metals, and the reader must not suppose that the methods we have sketched and symbolized are the only methods by which metals are taken from the circulation, brought to rest and finally aggregated in veins. We have sketched only the chief method. We have laid stress on the agency of life and death because, in some way, they are the factors by which all metals, in every case, are taken from the circulation of nature and *fixed*. Those who saw, in our Centennial Exposition, a fossil coal plant whose bark was filled with plumbic sulphide, or galena, must have seen that the decay of the plant was in some way instrumental in precipitating the lead. If they gave the subject anything more than a passing thought they would have seen that the lead must have been held in solution and precipitated into the tree during the process of decay. A soluble form of lead is the plumbic sulphate, Pl. $S_4$. Suppose that the molecules yielded their oxygen to the carbon of the decaying plant. They would be reduced to plumbic *sulphide*, Pl. S., which is insoluble. It would be precipitated and take the place of the carbon burned by the oxygen it had yielded.

In the sandstones of the Connecticut Valley and New Jersey we often find cupric sulphide encrusting fossil plants. The history of this copper must be similar to the history of the lead in the coal plant. The copper must have been soluble. A soluble form of copper is cupric sulphate. Cu. $S_4$. The molecules yield their oxygen to the decaying plant and become molecules of cupric sulphide, Cu. S., which are insoluble. In each of these cases the history of the metal is clearly indicated. We can see that if none of the plant had decayed no metal would have been precipitated, and that if *all* of the plant had decayed *more* of the metal might have been precipitated. But in that case no fossil would have been left to tell us how the metal came there.

the center long before *cooling* would cause it at the surface. From within *outward* is the growth of a solid world, and then from without *inward*, and the solid crust grows toward the solid core through a zone of molten lava.

After rock-making in the crust came ocean-making in the air. Condensation of the vapors occurred and the still glowing crust was enveloped in a dense and steaming ocean.

A cooling crust shrinks and this shrinking impresses on the face of a globe its first features. A mountain chain begins not in an upheaval but in a subsidence. Let the line *a b* represent the bed of an ocean.

Fig. 8.

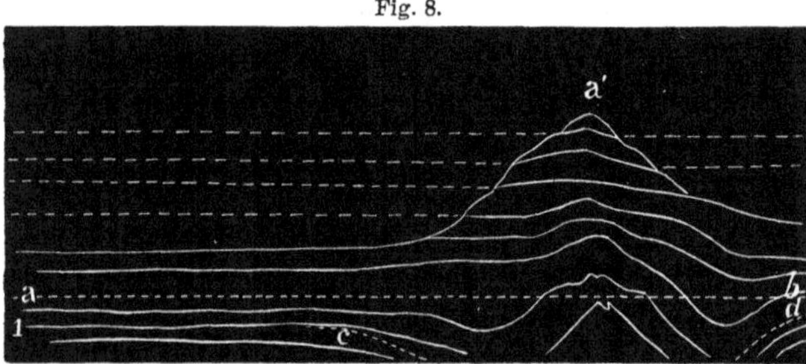


Through unequal cooling and contracting in the crust below, suppose the bed from *c* to *d* sinks to the curve represented by the dotted line. Sediments would accumulate in such a trough in greater thickness than elsewhere. The accumulation of sediments might keep pace with the downthrow and the bed of the sea remain everywhere nearly at the same level. Suppose that in such a trough sediments were to accumulate seven miles and three thousand feet thick. The

Fig. 8. An ideal section showing the process of mountain-making. Dotted lines above show the strata which have been eroded.

bottom might dip into the molten zone below and be melted off. Pressure of the superincumbent masses and pressure from the sides on a trough so weakened, would flex and crumple the lower rocks as your hand might crumple the leaves of a book. Suppose now that the bed of such an ocean becomes dry land. Suppose that at *a* the sediments which in the trough attain a thickness of seven miles and three thousand feet, are only four thousand feet thick. At *a* they are level. In the trough they are flexed and fractured. At *a* there would be a plane. Over the trough there would be a mountain. If the bed of rock represented by figure 1, which, in the plain, we suppose to lie at the sea-level, were to lie at the same level under the mountain, then the mountain would be forty thousand feet high. But if this rock-bed has been carried down thirty-six thousand feet, then the mountain would be only four thousand feet high.

By these suppositions we have read the history of the Appalachian Mountains. In the Valley of the Mississippi, as at *a* the strata are nearly horizontal and have a thickness of four thousand feet. In the Appalachians as at *a′* the strata are flexed and have a thickness of forty thousand feet. If these mountains are not seven miles and three thousand feet high it is because their base has been carried down about thirty-six thousand feet! That their rocks are folded and plicated is proof that they rested not on solid rocks but on a molten paste, as flexing would be impossible except over a yielding support. That they are not very high mountains is proof that they occupy an area of subsidence. That they are mountains at all is proof of the thinning out of the strata eastward and westward, and of mighty erosions by running water.

In the history of the Appalachians we have, in outline, the history of mountain-making everywhere and through all time.

Connected with the history of mountains is that of volcanoes.

To know what titanic forces still lurk in the earth's crust, or still are generated there, we have only to trace on a map the zones of volcanic activity. Along the coast of the Pacific from Fuegia to the Aleutian Isles; from the Aleutian to the Isles of Sunda; from Sunda westward through Sumbawa and Java and Sumatra; from Sumatra north-westward to the burning mountains of the Indian Archipelago, and from the Moluccas south-eastward through New Guinea to New Zealand—there is a belt of fire which girdles four continents, the two Americas and the external borders of Asia and Australia. From the Thian-Chan Mountains of Central Asia through Lake Aral, and the Caspian and the Caucasus and the Grecian Islands and Naples and Sicily to the Azores, extending over ninety degrees of longitude, is another volcanic train. Intersecting this at the Azores is still another train extending from Jan Mayen through Iceland, the Canaries, the Caribbean Islands, Ascension, St. Helena, to Tristan d'Acunha. So is the world zoned with burning mountains and islands.

Most volcanoes are intermittent, but not all. Rancagua rests neither day nor night but continually throws out ashes and vapor. Stromboli is forever belching forth smoke and flame and cinders. Villarica in Chili is never quiet. Popocatapetl which rises eighteen thousand feet above Mexico, is known to have smoked and burned without a day of rest since the conquest by Pizarro. Kirauea in the Sandwich Islands is a crescent-shaped gulf six miles in circumference and fifteen hundred feet deep, a gulf of lava, molten in places and in places cooled into a crust, but studded over all its face with cones that smoke continually and pour out now and then such masses of glowing lava that the whole gulf becomes a lake of liquid fire, terrible to behold! Seen in a dark night it is Milton's hell,

> "——forlorn and wild,
> The seat of desolation, void of light,
> Save what the glimmering of these livid flames
> Casts pale and dreadful."

The subterranean forces seem most titanic when their manifestations are paroxysmal. Vesuvius which is three thousand feet high has thrown scoria four thousand feet higher. Cotopaxi has thrown fragments of lava six thousand feet above his summit, which stands eighteen thousand feet above the ocean. Stromboli, which sometimes is paroxysmal, has hurled cinders ten thousand feet above his summit, and shot forth masses of white-hot lava which have globed themselves by rotation in the air and fallen as volcanic bombs fifteen feet in diameter. Coseguina, on the Gulf of Fonseca, has thrown his cinders over the whole breadth of Guatimala to Truxillo on the Gulf of Mexico.

We turn from living volcanoes to those which are dying.

Cinder and vapor eruptions are the dying efforts of a volcano. Kirauea ejects no cinders but only lava so hot that little jets are sometimes tossed up by the wind and spun out in fine glassy threads known by the natives as "*Pele's hair.*" Kirauea is in full vitality. Vesuvius throws out cinders chiefly, an indication that his vitality is waning. Mt. Kea, near Kirauea, is almost dead, having ceased to throw out lava and formed cinder cones on his summit.

From the dying we pass to the dead.

Chimborazo in Quito and Tacoza in Peru are dead. In Syria and Palestine there are dead volcanoes. Some of these may have been active within the memory of the Hebrews whose literature is draped in imagery borrowed from the volcano, "He toucheth the hills and they smoke." "A fire shall devour before him and it shall be tempestuous round about him." "The voice of the Lord divideth the flames of fire." "The voice of the Lord shaketh the wilderness; the Lord shaketh the wilderness of Kadesh." Dead volcanoes of an earlier age cover a large area in Central France. In the ashes thrown out by the extinct volcanoes of Auvergne ages ago are found the skeletons of extinct quadrupeds—an old event explained by a recent one. The ashes from Coseguina, in 1835, covered the earth twenty-four miles away, to the depth of twelve feet, suffocating and roasting thousands of cattle and wild animals and birds.

Still older volcanoes we can trace by their tracks of lava. A vast area along our Pacific slope was flooded by outpourings of lava which form ramparts of basaltic columns along the upper Columbia, the Table Mountains, and the lofty peaks of the Sierra. In the same geologic age immense overflows of lava buried a part of Scotland, Ireland, England, and France. In a still earlier age floods of lava were poured out along our Atlantic slope from Nova Scotia to North Carolina. The ridges of trap which skirt the Bay of Fundy; the Palisades along the Hudson, Bergen Hill in New Jersey, Mount Holyoke and Mount Tom in Massachusetts, and the "Hanging Hills" of Connecticut represent a portion of these early outflows. Now these ancient lavas are composed of silica 48, alumina 16, protoxide of iron 15, lime 9, magnesia 3, potash 2, and soda 2. We have omitted fractions which would bring our numbers up to 100. We have given the composition of this rock which is called *doleryte*, because of its great importance. About four-fifths of the rocks which have come up and do come up molten through the crust are *doleryte*. The lava of Kirauea is doleryte but where it cools rapidly it forms a black glass called obsidian. Doleryte is the most abundant, and, from the date of the earliest known outflows, has been the most abundant rock in the liquid zone under the earth's crust. If we may judge of what is left by what volcanoes have taken, four-fifths of that portion of the globe which is still molten is composed of silica, alumina, iron, lime, magnesia, and potash. And if we could strip off the super-crust we should find a globe of molten lava composed chiefly of these elements. We should find the world as it was before solidification began on the surface.

From the living volcanoes we have gone back through the dead to a time when the whole globe was a sort of volcano without its cap of rock. And the lava which cools to-day in Kirauea will give us the best picture of the first rocks which appeared on the globe's cooling surface. Cooling slowly that lava becomes a hard, compact, dark-colored doleryte. Cooling rapidly it becomes a black glass. Late experiments on molten rock-matter, made by Bischof, Deville, Delesse and Mallet, have shown that the *glass*-state is reached in consequence of rapid cooling and the *stone*-state in consequence of slow cooling, and that glass will become stone if melted and allowed to cool slowly. The original crust might have been black and glassy had the earth cooled rapidly. But Helmholtz has shown by calculations based on the rate of cooling lavas, that the earth must have been *three hundred and fifty million* years in cooling from the melting point of doleryte down to the boiling point of water. Cooling so slowly it crusted over, not in glass but in rock. We may learn patience. If the world had hurried up her cooling we might to-day have been walking over shards of glass. Having taken things rather leisurely it laid the foundations in honest rock. Volcanoes are outflows of the fire-sea, the residue of the primitive molten globe, which, age by age, has been retreating deeper and deeper under a thickening crust. Their history has a close connection with that of mountains.

We have seen that a mountain chain begins in an area of subsidence; that the heavier sediments accumulate in the trough, and that the lower rocks of this subsiding area, either dipping into the molten zone and melting off, or being weakened by intense heat, break down and open vents for the escape of the imprisoned lava. The principal volcanoes living now, as Hunt has shown, occupy areas of subsidence and thick accumulations of sediment. The chief volcanic outbreaks through all the geologic ages have occurred along subsiding areas weighted with heavy accumulations of sediment.

Intimately connected with volcanoes are earthquakes. The great volcanic zones are zones in which the crust is often shaken by earthquakes, but the tremblings are not limited to such areas.

On the first of November, 1755, a sound like that of thunder was heard under Lisbon. The elements above seemed in sympathy with the distress below. The face of the sun grew red and the air grew hazy. Within six minutes the greater part of Lisbon was in ruins and sixty thousand men, women, and children were perishing. Cintra, Julio, Estrella, and Arrabida, the greatest mountains of Portugal, were trembling from top to base, were opening in deep rents and fissures and emitting flames from their summits. A new marble quay at Lisbon went down, carrying a multitude of people. Many vessels anchored near it went down, and it is said that not a body carried down on the quay, and not the fragment of a vessel carried down from the harbor, ever rose to the surface. Eight hundred feet of ocean covers the site of the marble quay. Far out at sea ships received the shock as if they had struck on a reef. In the West Indies a tide of inky blackness rose twenty feet where the usual tides are two. The coast of Africa trembled almost as violently as the coast of Portugal, and a village of ten thousand people was swallowed up. The shock was felt in England, on the Alps, on the shores of the Baltic, on the lakes of Scotland, and on the far-off lakes of North America. Humboldt, in the first volume of Cosmos, estimates that a portion of the earth's surface more than four times as large as Europe was shaken by the tremor which destroyed Lisbon. Although this was the greatest earthquake ever known and recorded by man, it does not stand out on the last pages of the earth's history in complete isolation. An earthquake in Peru, on the 14th of August, 1868, sent its waves northward along the coast of the Pacific to Oregon, and over the Pacific to New Zealand and Australia. A great earthquake in Japan, in 1854, sent its waves across the Pacific to California and Oregon, where they were recorded by the self-registering tide-gauges. Within the present century a single earthquake in Chili added to the continent of South America a mass of rock estimated at fifty-five cubic miles. A series of earthquakes in 1811–12 subtracted from North America a volume of rocks not estimated in cubic miles. An area in the Mississippi Valley eighty miles long and thirty miles wide sank and has remained permanently at a lower level. A large portion of New Zealand has been lifted up from one to nine feet, while a large tract in the delta of the Indus has sunk. In the archipelago of New Zealand the crust has broken on a line about a hundred miles long. On one side of the line the surface has remained at rest, while on the other it has been pushed up about nine feet.

We look now from the uplifts and downthrows which have occurred within our own century to those recorded in the earlier ages of the globe's history. If the coast of Chili records an uplift of three feet within our own time, the fossils of the Andes record an uplift during a recent geologic period of many thousand feet. If in New Zealand we have the record of a recent fracture and uplift along one side, of nine feet, in the Pennine Hills of England the rocks record a very ancient fracture and an uplift on one side of a thousand feet; and in Pennsylvania they record a break and an upthrow along one side of *twenty thousand feet.* The Pennsylvania fault dates from the same remote past as the Pennine fault of England. The fractures and displacements wrought by earthquakes within the present century are the same in kind as those which have occurred on so grand a scale through the æons of the past. Quakings of the crust, so fearful and destructive now, were a thousand times more fearful in the remote past when neither man nor his works were here to destroy. Such a titanic force is that which has caused the earthquake. And why not titanic?

If we drop a molten sphere of silica and alkali into cold water, it will cool rapidly and crust over with glass. Such a sphere is called a Prince Rupert drop. The interior will cool and contract more slowly. In consequence of the slow cooling and contracting within, the glassy crust is under tension. Atom presses against atom and the tension is so great that a scratch on the surface, destroying the equilibrium, causes the little sphere to explode and fly into fragments. Now what is the earth but a great Prince Rupert drop let fall, not into water but into empty space, and crusted over, not in glass but in rock? The crust is under the same kind of tension, and for the same reason, as the glassy crust of the Rupert drop. In a quarry near Belfast, Maine, we have seen a stratum of granite suddenly expand and lift itself up; and we have seen a mass of granite, when the strain was partly removed, explode with a loud noise and fly off into fragments. The chisels of the quarrymen were like the scratch of a file on a Rupert drop. Prof. Niles has reported similar occurrences in a quarry at Monson, Massachusetts. The entire crust is under a strain like that of the leaves of granite in these quarries. Now any *local* change of temperature in the crust, any melting of rocks and freeing of their gases at the seat of volcanoes, destroys the equilibrium, as a scratch on a Rupert drop. Fractures and vibrations are the result. A vibration, beginning deep down on a line of fracture, is propagated through the crust and felt on the surface as an earthquake; for rocks are more or less elastic. The shock received by a ledge from a passing railway train may shake your window-panes a quarter of a mile away. If quarrymen were working under the track they would feel a slight earthquake with every passing train. Under portions of Paris there are stone quarries, and the rattling of carriages on the pavement above is felt in slight jars by the workmen below.

The great vibrations start from the seat of volcanoes, where the subterranean fires seem to be fed by water which has percolated from the ocean through rents and fissures. That such water does percolate the crust and mingle with the lava is shown by the salt and hydrogen ejected by volcanoes. That it did so percolate and so mix with lava eighteen hundred years ago is shown by the great multitude of silicious sea-shells in the lava which envelops Pompeii and Herculaneum. Now Bunsen has shown that as you heat iron red-hot and white-hot, so under great pressure *water can be heated red-hot and even white-hot.* Water that finds its way deep down to the roots of a volcano must be under enormous pressure, and red-hot or white-hot water mixed with lava would generate a force of which we can form a faint conception when we picture a mass of rock three hundred cubic feet in bulk shot right up through the throat of Cotopaxi, which is eighteen thousand feet long, and nine miles through the air beyond! This is not all. We are to remember that as water under great pressure may be white-hot and not break into gases, so gases under such pressure may lose their gaseous character and become fluid. Lava mixed with white-hot water and fluid gases would be a force of which we can form a very faint conception when we recall the shattering of deeply-buried rocks, and the shaking and uplifting of the incumbent strata when Lisbon was engulfed.

Thus features have been impressed on the cooling and shrinking globe by the reaction of its internal heat on its crust, and the erosion of its surface by water. When the globe was young and the crust thin, volcanic eruptions were at the minimum and earthquakes at the maximum. A thin and yielding crust would not break, and we find hardly a trace of volcanic activity recorded in the earlier chapters of the earth's history. But a thin and yielding crust has everywhere been flexed and plicated. Everywhere the young crust has recorded its own instability. In Nova Scotia it has recorded at least seventy-five oscillations of level—land emerging from the sea and again sinking under its waves, and these alternations repeated seventy-five times and registered in seventy-five beds of sediment accumulated under water, and seventy-five intercalating beds of coal, each the growth of a marsh. It is a fact of great significance that during all these movements of an unstable crust there were no fractures and no eruptions of lava. In after times the crust broke and the nether sea poured out floods of lava from Nova Scotia to North Carolina. In the period of mountain-making recorded in the Appalachians, while the crust, under great strain, was bending and here and there breaking and slipping down along the line of fracture, the outflows of lava were insignificant. In a far later period of mountain-making, recorded in the Sierras of Oregon and California, when the crust was thicker and firmer, it broke down and immense floods of lava were poured out over the surface. Earthquake activity culminated very early in the globe's history. Volcanic activity culminated, it would seem, during the geologic period which immediately antedates the present.

The reactions of the molten zone on the crust, whether in volcanoes or in earthquakes, are on the wane. The crust grows thicker, grows thicker as a crust of ice on a pond, by additions from the liquid zone to its nether surface. Heat from that zone is still radiating through the crust into outer space—heat enough every year, according to Sir William Thomson, to melt seven hundred and seventy-seven cubic miles of ice. At this rate, or *any* rate, that molten zone must eventually cool, and the solid crust meet the solid nucleus, and the whole globe become cold, solid, dead. Mt. Kea, which is dying, will soon die and be numbered with the burned-out mountains around it. Vesuvius, which already "pales his ineffectual fires," will join Chimborazo and be counted with the dead. One by one the burning mountains will be quenched, one by one the living pass away and swell the long list of the dead.

Our sister globes will help us to complete the story of our own. For all the spheres are of the same essence and under the same law.

In the concert-room you know what keys of a piano are struck by the notes which fall on your ear. And the note, you have learned, is simply a pulse of air set in motion by the vibration of the wire. If now you could strike, let us say, an atom of copper as you strike the tine of a tuning-fork or the key of a piano, you would throw that into vibrations. The air could not take up and repeat these vibrations as it does those of the tine or the wire. If you could take away all the air between a burning gas-jet and your eyes, the jet would still shine for you. Something which is not air carries the light and the heat. It is a finer and an all-pervasive air called *ether*. It would feel and take up and bear along the vibrations of a struck atom. But you cannot strike an atom with a hammer less delicate than an atom of oxygen. The quaver of the atom so struck is borne to you in waves of ether which your senses translate, the longer ones into beat, and the shorter ones into light. A flood of air-pulses falling on your ear gives you *noise*. A flood of ether-pulses, of different lengths, falling on your eye gives you white light. Air-pulses of one length and beat, falling on your ear, give you a certain pitch or note. Ether pulses of one length, falling on your eye, give you a certain color. Colorless light is analogous to *noise*. Colored light is analogous to *pitch*. Now the quaver from an atom of copper is of a certain length, and when it beats against your eye your brain translates it into *green*. Green is the note poured out into ether by burning copper. Let it be known that each element has its own beat and it will be seen that as each vibrating wire of a piano reveals itself to the ear in its pitch of sound, so each burning element reveals itself to the eye in its pitch of color. And let it be known that the ether is cosmic and pervades all things and fills all space, and it will be seen that if copper or any other element known to us is burning in the sun, its quaver will be borne to us on the ether and its presence in the sun will be published on the earth by its color-note. The spectroscope is an instrument which resolves the colorless light, "the noise," of the sun into its color-notes. It has shown us that the same elements which compose the earth compose all the universe revealed to us by its light. The spheres, being of the same essence, must be under the same law.

"In tracts of fluent heat" all spheres began. Through times commensurate with its bulk, each sphere runs through its geologic history, cooling and shrinking and developing under its shrinking crust the same titanic forces which have shaken the earth and girt it with trains of burning mountains. All spheres may not safely carry, as the earth has carried, the terrible forces their own evolution develops within. As many a fruit carries an enemy in its core, and withers and falls at last from the bough, so many a world, a prey to the forces at its core, may have fallen and crumbled back into its elements. Many a world may have exploded like a bomb, or broken into fragments like a Rupert drop.

The smaller the sphere the more rapidly it runs

through its geologic changes, the more rapidly the nether fire-sea cools, and the sooner the solid crust meets the solid nucleus and nature brings up in rest.

The face of the moon is a picture of rest, and at the same time a picture of spent forces. Mountains and "ring mountains," and lava plains and vast chasms, and high craters and deep volcanic throats, but no volcanoes, no moonquakes, no oceans, no life, no motion — that is a picture of the moon. Everywhere is the fixedness of death. In the moon nature is at rest. But there is no "*natura*" there, no "continuous on-flow." The moon then is a product of spent forces.

The moon is a sphere in the last stage of its evolution. When the fires of the earth shall have died out and there shall be no more reaction between surface and interior, then must the face of the earth become even as the face of the moon.

In rocks, we have said, nature is at rest, and we have seen by what antecedent activities she came to her rest in the rock and in the vein-stored metal. These activities were molecular and vital.

We have seen that when rock-masses were formed, other and more stupendous forces acted upon them and shattered them and pierced them with melted lava, and pressed them into wrinkles and lifted them up into continents and pulled them down under oceans. These forces are to the mass as the molecular forces to the particles, and when *they* are spent nature comes to rest alike in mass and particle. Such rest is death, world-death.

This may not be the end. The analogy between a world and an organism may not end with death.

Decay may follow death in the one as well as the other. Munier, the French Academician, has labored to show that in a world which is cold to the core; which, by absorption into the crust, has lost its oceans and its air; which is dead thermally and electrically; atom will lose its hold upon atom, and molecule upon molecule, and the dead world like a dead body will be resolved into the elements whence it came. It may be so; probably it is so. Then as the body you wear will yield its elements to be organized into other bodies and so live on in body after body through all the coming ages of life, so a dead world will crumble back into atoms to be globed again into other worlds, and so cycle on in world after world through all the coming ages of eternity. And as the matter of your body has been organized in bodies that lived before you, so the constituents of our earth may have been nebula globed into earths which accomplished their cycles and melted back into nebula, and so may have cycled on through nebula to world and world to nebula through the deeps of the past eternity.

"Through the vastness arching all,
I see the great stars rise and fall,
The rounding seasons come and go
The tidal oceans ebb and flow;
The tokens of a central force,
O'erlap and move the universe;
The workings of the law whence springs
The rhythmic harmony of things,
Which shapes in earth the darkling spar,
And orbs in heaven the morning star."

## CHAPTER II.

Under the Seas — Rock-Making about the Poles, along the Shores, in the Great Deeps — Petrifactions — Fossils "the Parents of Rocks" — Types of Life — The Present an Outgrowth of the Past — Bird's Flight View of the History of Life as read from the Rocks of North America.

THE earth has written its own history. If we could cleave it to the core and then go down along the cleft we might pass first through clay or sand or gravel. Under this top-dressing we would find rocks, either crystalline or fragmental. Crystalline rocks are sometimes called *massive*, and fragmental rocks *stratified*. The crystalline are, in general, deepest and oldest. The fragmental, in general, lie nearer the surface and are younger. How were they formed? To answer, we have only to look about us.

Rain and frost and wind-blown sand are slowly pounding the rocks into dust. A brook which trickles from a mountain brings down spoils which the elements have chipped from the rocks. Freighted with these spoils, the brook itself becomes an agent of destruction. For as grains of sand when dashed by the wind against a ledge of granite, pound it into dust, so when borne on water they wear away the rocks against which the current throws them. A stream would plow for itself a channel straight down through the top-dressing into the rocks, but rain and frost and melting snow wear down the banks, and thus river erosion and bank erosion sculpture the face of the earth into beauties of hill and valley.

Southwest of Utah is a land threaded by rivers which are fed by rain and snow from the slopes of the Sierra. In that land there is not rain enough or snow enough for bank erosion, and the rivers plow their channels straight down into the rocks. On the brink of a gorge which you can almost leap across you look down along a wall of rock in places more than four thousand feet, on the river which flowed, ages ago, at the level on which you stand.

Fig. 9.

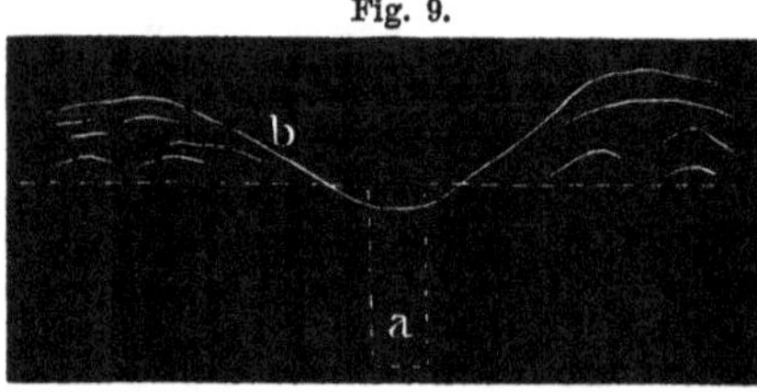


Now the rock taken from the river valleys and river cañons—where is it?

We stand by the sea near the mouth of a river. Along the beach we see a fringe of sand or gravel. It is spoils from the land thrown down by the sea. But the finer sediments brought down by rivers are diffused through the ocean and do not fall to the bottom until they have passed through the mill of life. In every drop of water on the surface of the Polar Sea are little plants known under the name of *Diatoms*. (Fig. 10—*a*.) These plants take from the water atoms of silica and build them into cases shaped, some of them like a boat, and sculptured, all of them, in lines of beauty not matched by nature where she works her patterns on a larger scale. At death these sculptured cases fall to the bottom. The

Fig. 9. Ideal section of Canon and Valley.

dredge brings up from the bed of the Polar Sea little else than white shells of Diatoms.

Fig. 10.

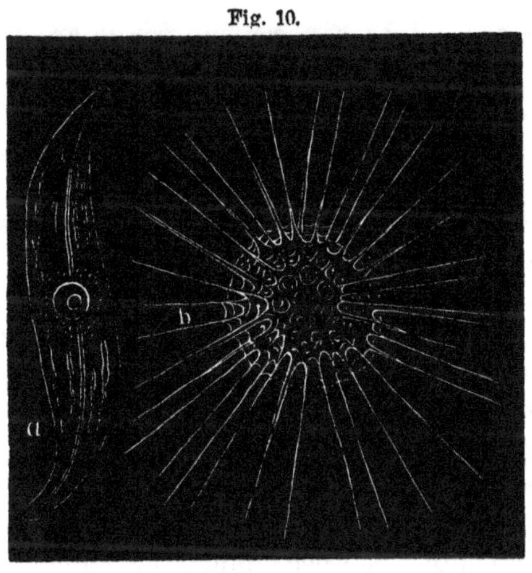


Between the polar zones the face of the sea swarms with the lowest of organisms. The Amœba is merely a bleb of jelly, a body without members, an organism without organs. Such specks of jelly, no mouth or *all* mouth, no stomach or *all* stomach, no limbs or *all* limbs, take from the sea atoms of lime and build them into shells of almost every shape the geometer can fancy. Our plate represents one of these shells dredged from the deep-sea bottom by the naturalists of the "Challenger." The shell is globular in form, clear and transparent, and is penetrated by a great many little pores, each pore lying at the bottom of an hexagonal pit. At each angle of the hexagon the crest gives off a delicate calcareous spine.

Fig. 10. Globigerina *b*, and Diatom *a*.

The shell is chambered and the chambers are filled by the orange-colored jelly of the "Rhizopod," in English, the *Root-foot.* Part of the body exudes through the pores, spreads out as a flocculent fringe around the shell, and shoots forth into long, streaming "pseudopodia," or *false-feet.* From the appearance of the pseudopodia a new name has been given to the shell-bearing Amœba. They are called "Ray-streamers." This kind of Ray-streamer is called "Globigerina."

As the Diatoms take silex from the sea, so the Ray-streamers, by their vital chemistries, take lime, and after death, by the chemistries of decay, take silex and alumina and iron and potash. These elements, uniting and forming a green mineral called *glauconite,* filter through the pores of the shell and make a glassy cast which falls to the sea-bottom. If we put the dredge down where the depth is between one and three hundred fathoms we shall bring up green sand. And if we examine a grain of this sand under the glass we shall find it the cast of a Ray-streamer's shell.

If we sink the dredge where the depth is between three hundred and two thousand fathoms we shall bring up a cream-colored ooze, not, as the *green* ooze, the *casts* of shells, but shells themselves. This is called "Globigerina ooze."

If we put the dredge down into the abysses of ocean, three miles or more, it brings from the bottom *red* ooze. Not a shell and not the cast of a shell can be found in this deepest ooze. But Wyville Thompson and his colleagues of the Challenger have investigated this red clay which their dredge has brought up from all depths below three miles, and they have been led to the belief that it is the residuum left after the lime of the Globigerina ooze has been dissolved away. According to this view the very clay of the deep sea-bottom is the product of life, and the ocean, everywhere except along the shore where it throws down a fringe of sand, holds the sediment brought to it by the rivers and gives it up only to its living things or its decaying dead. Passing through the laboratory of life, the sediments from the land come to rest on the floor of the sea. There they await transformation into rocks.

The rocks along the shores of the Mediterranean at Zante and Oran; in Bohemia at Bilan, and in Germany along the Rhine, are made up of the flinty cases of Diatoms like those which are lying now as beds of snow on the floor of the polar sea. The green sand of New Jersey and England is made chiefly of the casts of Rhizopod shells like those which form the bed of the ocean at depths not less than one hundred or more than three hundred fathoms. Chalk and white marble are made of Rhizopod shells like those which form the ooze of the sea-bed at depths between three hundred and two thousand fathoms. Slate and shale are hardened clay like that which as a reddish ooze lies in the abysmal deeps. Late experiments communicated to the French Academy show that common glass when exposed to a temperature of 400° C., in presence of its own volume of water, swells up and crystallizes. Late investigations by Sterry Hunt have shown that feldspar and mica can be crystallized from hot alkaline solutions in sealed tubes. Other investigations by Daubré have shown that alkaline solutions at temperatures above 700° F.

combine with clay and form feldspar and mica. The concurrent testimony of these experiments is that crystalline rocks can be formed out of fragmental rocks, or sediments, by heat and water impregnated with alkali. The heat may be generated in the lower mass of the sediment by pressure of the superincumbent masses. The water is furnished by the ocean. And now, as granite and all rocks of the granite family may result from changes wrought on sediment by alkaline water and heat, it may be that all that portion of the earth's crust ever exposed to the eyes of men has been again and again through the mill of life.

Under the air the world is wasting away. Under the sea it is building anew. The new is created out of the ruins of the old. Nothing is lost. Between old rocks and rocks not so old the go-between has been ocean and life.

"Fossils," said Linnæus, anticipating a hundred years of science, "are the parents, not the children of rocks." *Microscopic* fossils, to speak in the vein of Linnæus, are the parents, fossils not microscopic are the children of rocks.

In the ocean an animal dies, sinks to the bottom and lies imbedded in the ooze. Little by little atoms of mineral replace the animal matter in the bones, the teeth, or the scales, and these hard parts become petrified. If we could explore the bed of the ocean we would find, strewn through it, the remains of all bone or shell or scale-bearing animals which have lived in the waters above or on the mud below. Land animals, too, we would find, for as the land pays to the sea its tribute of mud, so it pays its tribute of death. Even man goes down and the works of man. From England alone five hundred ships go the bottom every year. Our ships, our cannons, our coins, our bodies, entombed in the sea-mud, may be preserved forever. That mud, becoming rock, may be lifted up into a continent, and science may read in the strata which

Fig. 11.

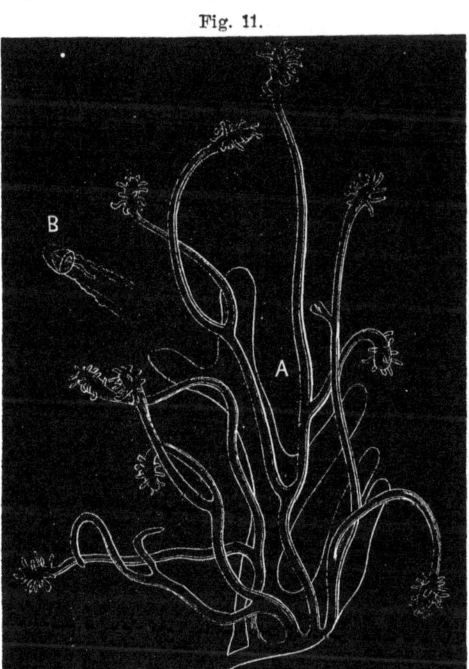


are *yet to be*, the history of our times. It is her boon to read in the strata which *are*, the history of all earth-times before us.

Some one has said that whatever lifts the mind out of the present and throws it into the far past or the far future elevates and humanizes. The past we are

Fig. 11. Coryne mirabilis.

now to traverse is only less than infinite. The light which illuminates it must be borrowed from the present.

If we look into a tranquil sea on any summer day we find animals without bone or shell, mere bodies of

Fig. 12.

jelly shaped like a saucer, or globe, or bell. If the jelly is bell-like or saucer-like, frills depend from the rim, and a tube, which encloses the stomach and terminates in a mouth, hangs from the hollow like a bell's clapper. Such jellies are called "Medusæ." Certain Medusæ produce progeny in which the animal is masked under

Fig. 12. Tubularia indivisa.

the form of a flower. Such an animal-flower is called a "Hydroid." In Fig. 11 a Hydroid-medusa is represented at B., and the Hydroid from which it came at A. The Hydroid is a hollow, branching stem, and each branch is tipped with a club-like expansion supporting tentacles.

The Hydroid-medusa, although seemingly an animal roving and killing and eating at will, is only the egg-bearing organ of a Hydroid. We give in evidence another Hydroid whose stem is crowned with a head which bears a wreath of tentacles, giving it the form of a delicate flower. Now the organ which develops from the head of *Coryne* into a bell-shaped Medusa with pendant frills and mouth, develops also from the head of Tubularia from which it is never freed. After its eggs have been cast into the sea it withers and dies.

Fig. 13.

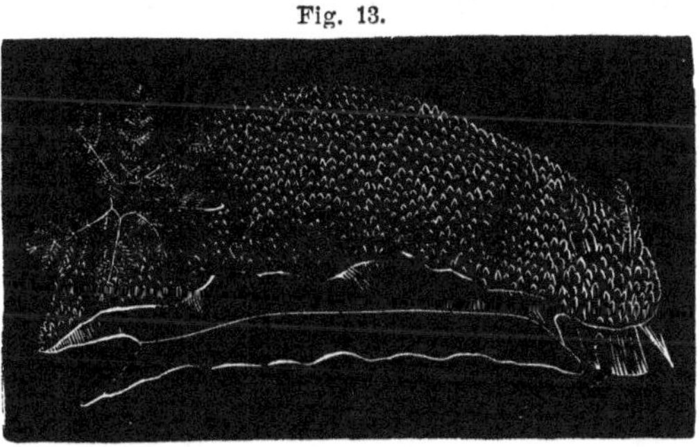

In the sea are soft, sac-like bodies, naked or encased in shells. Fig. 13 gives an illustration of a naked mollusk. The edges of the mantle are seen below, two frond-like antennæ appear on the top of

Fig. 13. Doris plumulata.

the head, and thick-set papillæ extend along the back to the other extremity, where the lung appears like a tuft of delicate sea-weed wrought into an eight-rayed star.

Fig. 14.

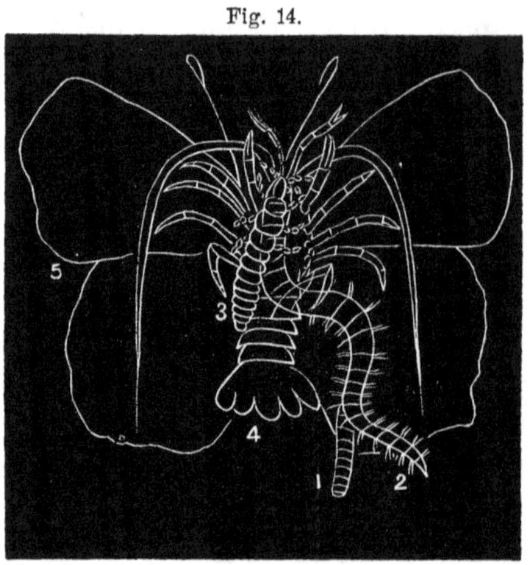


In the earth is an animal whose elongated body is made of segments, each segment a mere repetition of every other. To the segments of the Earth-worm (No. 1 of Fig. 14) add bristle-like appendages, and you have the Lob-worm of the sea (No. 2). Let the body segments differ, one from another, and let the appendages be jointed, and we have an Insect (No. 3). Let the differences between segment and segment be intensified, let the animal breathe not through wind-pipes but through gills, and we have a Crustacean (No. 4). The ground-pattern is the segmented Earth-worm. Differentiation of the segments, and the appendages from the segments developing into mouth-parts, or legs, or

Fig. 14. Articulate Types.—1. The Earth-worm. 2. The Lob-worm. 3. The Insect. 4. The Lobster. 5. The Butterfly.

wings, give rise to all the wealth of segmented species from Worms to Butterflies, and from Worms to Lobsters.

From the animal with a segmented body we turn to the animal with a segmented *axis* of the body.

The egg, as seen by the eye of science, is a globular cell with albumen at one pole and oil at the other. *The lowest microscopic animals are eggs through life.* The albumen which appears as a light spot in the simplest egg, in the egg of the higher animal appears as a germinal dot. The oil-mass expands and becomes known as the yolk. A groove is sunk around the equator of the egg, cutting through the yolk and dividing it into two parts. Another groove is sunk from pole to pole, dividing the yolk into four. Other grooves follow until the yolk is cut into microscopic cells.

So far the history of creation is the same for all the patterns of life. From this point, the animal with a segmented axis is carried up along lines peculiar to itself. The cells arrange themselves in three layers. From the outer layer is to come the outer skin, the spinal marrow, and the brain. From the inner layer is to come the lungs, the glands, and the delicate skin which lines the cavities of the body. From the middle layer is to come the muscles, the bones, and the blood vessels.

On the surface of the egg there appears now a delicate furrow. Along the furrow, on either side, the outer cell layer rises in the form of a ridge. The ridges meet over the furrow and grow together, forming the marrow-tube. If creation were arrested at this stage we would have the headless, brainless

Lancelet represented at *a*, Fig. 15. The tube, at first, is pointed at both ends. Head and tail begin alike. But development goes on and soon the head gets ahead of the tail. The fore-end dilates and changes into a roundish bladder, the beginning of a brain.

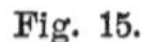

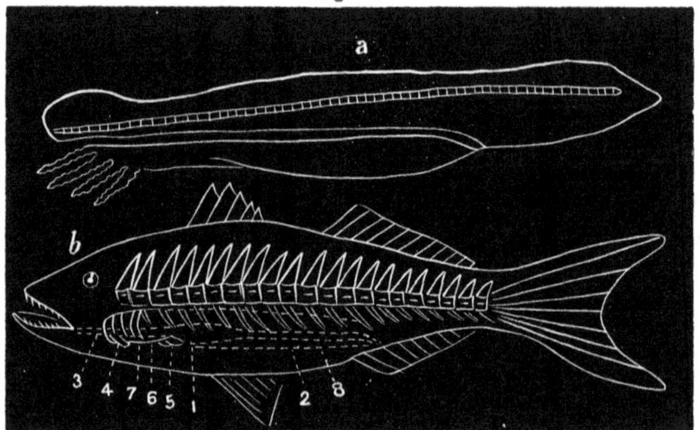


The stomach begins before the heart. Below the marrow-tube another tube is formed. An expanded portion of this tube is the stomach (1), a contracted portion beyond the stomach is the intestine (2), a straight, or slightly curved section before the stomach is the gullet (3), and a hollow bud from the gullet (4) is the beginning of a lung. In the Fish the lung is rudimental and is known as the swim-bladder.

The heart (5) comes in two chambers. In heart, the *whole* of a Fish corresponds to the *half* of a man. A great blood vessel (6) is carried forward from the

Fig. 15. DEVELOPMENT OF A VERTEBRATE.—*a*. A Lancelet. *b*. A Typical Fish. The segmented cord of Lancelet has developed into a vertebral column. 1. The Stomach. 2. Intestine. 3. Gullet. 5. Heart. 6. Great Blood-vessel Leading to Gills. 7. Gill-arches. 4. Swim-bladder.

heart to the gill where it divides into a number of branches (7) which curve upward, give the blood to the gill-tufts and then unite and form a vessel (8) which is carried back on a line nearly parallel with the marrow-tube.

The forth-coming animal must be placed in relations with the world around it. As lungs begin in buds from the throat, so eyes begin in blisters on the face. And as the true lung in a Fish never gets beyond the bud-stage, so in some of the Lobsters the eye never gets beyond the blister-stage.

In a Fish the senses are guides to the mouth or sentinels to report the approach of other mouths.

The marrow-tube and organ below it must have protection and support. Under this tube is laid down an axis of drum-shaped bones — the segments hollow at both ends if for a Fish, round at one end and hollow at the other if for a Reptile, and flat at both ends if for a Bird or Mammal. The marrow-tube is embraced by arches of bone which ascend from this axis, and the organs below are protected by ribs which descend from it — thread-like and flexible if for a Fish, rudimental if for a Frog, flat, inflexible, and united into a casque if for a Turtle.

The distinctive character of a Fish is the *limb.* It appears in every member of the order as a fin which corresponds imperfectly to the limb of a higher vertebrate.

In the Fish we have the ground pattern of all vertebrate animals, living or extinct. We have in this pattern "the promise and potency" of all forms of vertebrate life, as we have in the Earth-worm, potentially, all the forms of articulate life.

The history of creation is nothing to us unless its methods can be traversed by our reason. The past of the planet has no interest to us unless it holds the root of the present. The methods of nature *can* be traversed by the minds of men, and in the past were the seeds of the present. The Protozoan, represented in the Ray-streamer, the Cœlenterate, represented in the Medusa, the Articulate, the Molluscan, and the Vertebrate—on these patterns creation has woven the life that robes the earth to-day, and the minds of men may look in on her loom, see the play of her shuttle, and trace the strands of the robe back, changed but unbroken, into the cast-off life-robe of æons gone by.

We are to take now a bird's flight view back through the history of life as we may read it in the rocks of our continent.

We start from Florida. The rocks of this peninsula are limestone and the youngest of the continent. To see how they were made we have only to look into the bed of the Atlantic off the Florida coast.

A Polyp, which is a Cœlenterate animal in shape of a flower, discharges eggs from its mouth or buds forth other Polyps from its disc or sides. The egg, after roaming for a while through the sea, grows into a little Polyp and fixes itself to the bottom. The lime which a Polyp secretes from the water is placed atom to atom along the gelatinous plates which radiate from the stomach to the surface of the body. The skeleton of a Polyp is called *coral*, and, according to the mode of the Polyp's budding, the coral grows up into domes, or hillocks, or cups, or clustered columns, or leaves, or ribbons, or fans, or branching trees, or feathered moss. In domes of Astrea or branching trees of Madrepore, it is only the surface that lives. Older Polyps die and the new ones, budding from their sides, take their place and on their parents' skeletons build their own.

There is no verdure in the coral grove, but cup and dome and tree are in perpetual bloom. The grove is not like a garden where all beautiful things grow side by side. Sandy plains stretch from forest to forest and little saharas of the sea lie close to perennial blossom as delicate in form and rich in color as any that spread their petals on the air.

In the coral grove every nook,.cove, grotto, has its tenants; every "coign of vantage" its "procreant cradle."

"Life in rare and beautiful forms,
Is sporting amid those bowers of stone,
And is safe while the wrathful spirit of storms,
Has made the top of the wave his own."

Not very safe in their bowers of stone are the forms of life, rare and beautiful, while storms are raging above. Wrathfully sweep the storms over sea and land. Groves are uprooted on the land and coral groves are crushed in the sea. When a storm is on the deep the waters down through many a fathom share the rage and roll and dash against the coral. Boughs are torn away, stems are uprooted, domes and cups and trees are ground to powder. The abraded fragments fill up the pores in each coral stalk and the openings between branch and branch or tree and tree, and thus the grove becomes a reef of solid rock, such lime-rock as we find in Florida.

So Florida was built up from the sea, atom by atom, by the lives of Polyps and the abrasion of storms. And if we take the rate at which coral is growing now as an approximation to the rate of growth in ages past, Florida cannot be less than seventy-five thousand years old. But the same species which built the peninsula out of lime-atoms are to-day building the Keys off its coast. Seventy-five thousand years have registered no perceptible change in the life-forms of the Atlantic.*

From Florida we pass into Alabama and find a belt of white and bluish marl, composed of clay and lime and holding fragments of sea-shells. It is one of the upper members of a system of strata called "Eocene," *dawn*, for it registers the dawn of the present order of things. The marls of Alabama will take us back in the history of the earth further than the coral reefs of Florida. They hold the remains of a great sea-monster, in the words of Owen, "one of the most extraordinary mammals which the revolutions of the globe have blotted from the page of life."

The skull was long and narrow between the eyes; the nasal bones well developed; molar teeth had two fangs and serrated, triangular crowns; the parietal bones (1) (walls of the skull) were united atop in a

* Let the reader look now at a map of the globe. Two hundred and ninety coral islands he will see mapped in the Pacific ocean. Let him fix his eye on that part of the Pacific most thickly studded with islands, and apply the scale to it. He will find it four thousand miles long and six hundred miles wide. Coral groves flourished, or in by-gone ages did flourish, under all that immense area of ocean. Nor is this all. A single reef, broken here and there, stretches along the east coast of New Holland for a distance of a thousand miles.

suture. These are characters which place the animal in relationship with the Seals. The skull of a whale is broad between the eyes; the nasal bones are short and sometimes reduced to mere rudiments; the teeth — as in the Sperm Whale — have but one root; the parietal bones are small and do not unite by a suture as in other Mammals, another bone being interposed between them. The jaw of the Whale is drawn out like the beak of a Bird, a character we find in this extinct animal.

The shoulder-blade was whale-like in having the spine (2) appear as a ridge along the anterior edge. Articulating with it was an upper arm-bone, *not* whale-like as it presented a surface for a free joint to the lower arm-bones.

The larger vertebræ are eighteen inches long and a foot in diameter. Their centra are flat at the ends as in all Mammals, and large in proportion to their arches, as in Whales. The ribs did not unite with the vertebræ directly but by ligaments, as in Whales. The hind-limb seems to have been completely aborted, as it is in the Whale.

Fig. 16.

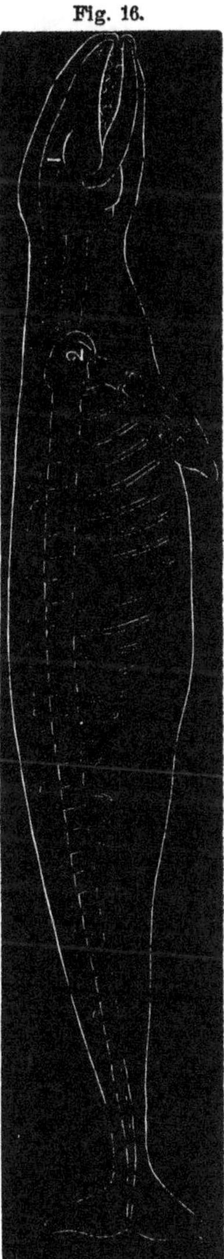


Fig. 16. Zeuglodon cetoides.

Balancing these characters, our Eocene sea monster seems to have been, in head, predominantly Seal; in spinal column and limbs, predominantly Whale. It was whale-like in form and bulk, having attained a length of nearly a hundred feet.

From the form of its tooth it has taken the name "Zeuglodon," in English, *Yoke-tooth.*

While Alabama was under an ocean swarming with Zeuglodons, prophetic of Whales, the region of the Rocky Mountains, having lately emerged from the sea, was dotted over with fresh water lakes. In Wyoming there is an immense basin stretching from the Uintah Mountains on the South far away to the Wind River chain on the North. In Eocene days that basin was the site of a lake as large as Lake Superior in our days. The lake stood there so long that the sediment brought down by rivers and spread out over its bottom grew up into strata eighteen thousand feet thick! The lake has been dry so long that rivers have plowed their cañons down through the strata thousands of feet deep! The rivers have been dead so long that no trace can be found of source or mouth! Rains have fallen and rills have eroded and sculptured the clay into domes and pillars and pyramids and spires — the semblance of vast ruins. It is a land of utter desolation. Stunted sage fringes the gorges, but hardly an owl sings its watch song from the lonely towers, or a spider weaves its web on the silent turrets, or a bat flaps its wings over the awful solitudes. Such, from the Uintah to the Wind River Mountains, is the world to-day. What was it in days gone by?

We take from these lake-strata the remains of a land Mammal with a strange assemblage of characters.

Fig. 17 represents a skull very long and narrow; facial parts very strongly pronounced; three sets of horn-cores, one rising from the nasal bones (1), one from the maxillary (2), and one from the parietal (3); an enormous crest above the brain-case; huge descending canine tusks (4) in the upper, and small molars

Fig. 17.

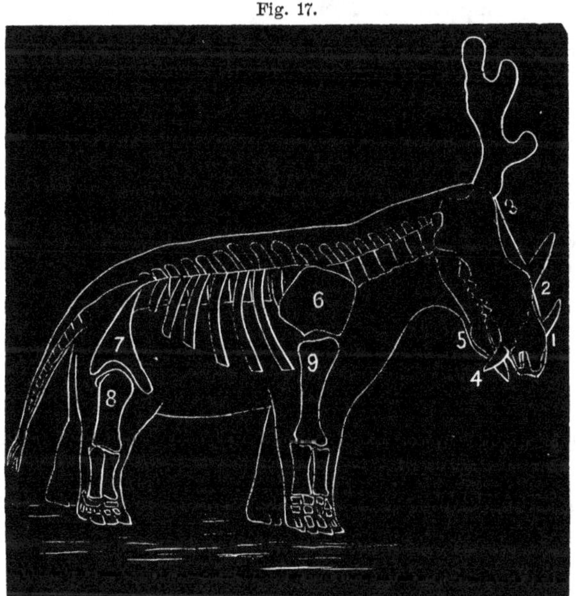


with slender roots in each jaw; small under jaw supporting a massive underhanging bony process (5). The brain-case shows the brain to have been small, exceedingly small, smaller than that of certain Reptiles, and in structure resembling that of a marsupial.

The shoulder-blade (6), almost triangular, had a large posterior expansion as in the Elephant.

Fig. 17. Uintahtherium giganteum.

The expansion of the pelvis (7) was four and a half feet, and in the general form of its elements it resembled that of an Elephant. The thigh-bone (8) was a third shorter than in the Elephant; its head was globular and its distal (lower) end was flattened. The tibia and fibula were distinct through their entire course, as in Elephant and Man, and other old-fashioned animals. The hind foot, short and round, terminated in five short hoof-bearing toes, a distinguishing feature of the Elephant. The articulation of the joints was flat, indicating but little power of motion.

The humerus (9) (upper bone of the fore-limb) was short and massive and in the main like that of an Elephant. In the Elephant the shaft of the radius crosses that of the ulna (lower bones of the fore-limb) obliquely, but in the animal we are reconstructing this feature is not so marked. The fore-foot had eight carpel bones, an interlocking series, as in the Rhinoceros and Tapir.

These details have been tedious but they will enlarge our intellectual horizon. We are as astronomers taking the parallax of a distant star. The animal which roamed along the banks of that Wyoming lake is as far from us in time as the star in space. What is its parallax? what its place in the scheme of creation? The long, narrow head with skull elevated behind into a great crest, the molar teeth and arch of the cheek-bone are characters which indicate relationship with the Rhinoceros. The absence of teeth in the premaxillaries points to the Ruminants. The vertical motion of the jaw points *away* from the Ruminants, toward the Carnivores.

The great expansion of the pelvis, the complete radius and ulna, the shoulder-blade and the hind-foot, are characters which affiliate our animal with the Elephant. The diminutive brain would place it with the pouched Opossum.

*The horns on the maxillaries, the concavity of the crown, and the enormous side crests of the cranium are characters which remove the animal from all living types.* We have found the ruins of an animal composed of Elephant, Rhinoceros, Ruminant, Marsupial, and a something unknown to the world of the living. Drawing an outline around the skeleton, we have a long head with little horns on the nose, conical horns over the eyes, and palmate horns over the ears, with pillar-like limbs supporting a massive body nearly eight feet high in the withers and six in the rump. Eliminating from its structure all that relates to the living, so many and such dominant structures remain that we cannot choose for it a name from existing orders. We will call it *Uintahtherium*,* which means, the Beast of the Uintah Mountains. Its companions were tapir-like, camel-like, hog-like, bear-like, tiger-like, and horse-like. A land of blight now, this Uintah region was once the fruitful womb whence issued the Tapir, the Rhinoceros, the Tiger, the Pig, the Horse, and the Camels which passed, one form into South America to appear in after ages as the Lama, and another form into Asia, there to bide the coming of man, and by serving him to civilize him.

Underlying these fresh-water strata is a system of marine strata called from *creta*, which means chalk,

* Uintahtherium of Leidy, Loxolophodon of Cope, and Dinoceras mirabilis of Marsh.

*Cretaceous.* The cretaceous rocks are green-sand, composed chiefly of casts of Rhizopod shells, and chalk, which is made almost entirely of Rhizopod shells. In America the system is represented by green sand, rotten limestone, and compact limestone, but no chalk except a little patch in Western Kansas. The system occurs here and there along the Atlantic border, from New Jersey to South Carolina. It has an enormous development in the Southwest. The cretaceous rocks extend from Texas north to the head waters of the Missouri, and thence westward through Dakota, Wyoming, Utah and Colorado, into California, and along the foot-hills of the Sierra Nevada to where,

> "Lonely as God, and white as the moon,
> Stands Mount Shasta."

They lie along the slopes of the Rocky Mountains, and were mud on the bottom of the sea, æons before these mountains were born. The cretaceous epoch was removed from the Eocene further, perhaps, than the Eocene is removed from the present.

In the ancient lake basin of Wyoming we find along the Black Butte, on Bitter Creek, upturned leaves of shale and coal. They were the floor on which the lake sediments were spread. They register the age in which the region of the Rocky Mountains, emerging from the sea, was lowland and marsh and lake. The shale is cretaceous, as its fossil sea-shells testify.

We take from these strata a vertebra measuring four feet four inches from the tip of the spine to the center of the body. It is slightly concave at both ends — a mere hint of fish-structure. Vertebræ belonging to the neck are slightly concave behind — a mammal structure. The hip-bone entombed with these vertebræ measures four feet from front to rear along the edges. It is more wonderful in form than in size. The ilium (upper part of the hip-bone) extends in front of the cup which receives the thigh-bone — a bird-character. The ischium is greatly elongated — another bird-character. In Reptiles the pubes incline

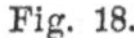
Fig. 18.

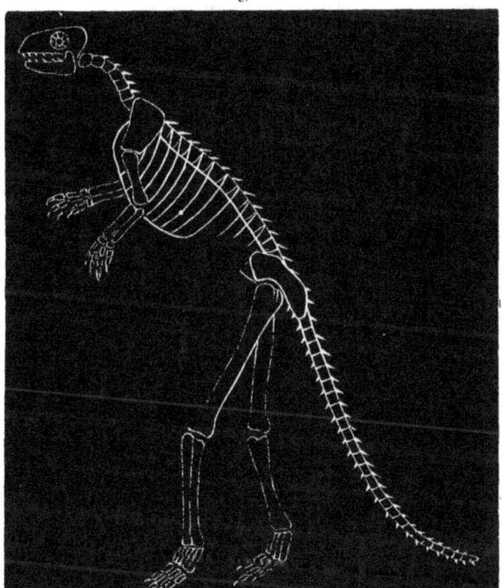

forward and downward. In Birds and in this Cretaceous monster they incline *backward* and downward. The thigh-bone is more than three feet long. The foot is bird-like, the tail large and reptile-like. The fore-limb was short and weak, the head small and in structure both reptilian and ornithic. Drawing around this skeleton the form the animal must have

Fig 18. Hadrosaurus.

worn, we have a huge nondescript about thirty-five feet long. Mounted on legs about six feet long, it waded through shallow waters or prowled over marshes, and was a something neither Bird nor Reptile nor Mammal, but compounded of all three. We shall meet it again, and many of its kindred, and hope for better acquaintance.

For this Dinosaur* shared the world with other Saurians similar in form and structure to itself, Saurians which left their remains in the Cretaceous marl of New Jersey; with Enaliosaurs† fifty feet long in the body and twenty in the neck; with Mosasaurs‡ eighty feet long, covered with bony plates and mounted on whale-like paddles; with Pterosaurs§ which spread their bat-like wings, twenty-five feet in expanse, over the site of Kansas, and left their ruins in the sandstone formed in the bays of her Cretaceous sea.

The uplift of the Sierra Nevada, the Wasatch, and the Uintah, registers, in the far West, the transition from an older geologic epoch to that of the Cretaceous. Rocks of the same age as the Swiss Jura, and hence called *Jurassic*, are borne up along the flanks of the Sierra, while beds of Cretaceous overlie them unconformably, showing that the Jurassic strata were formed before the chain rose, and the Cretaceous after it had risen. As the Cretaceous underlies the Eocene, so the Jurassic underlies the Cretaceous. Going *up* along the Sierra is going *down* in the earth's crust, for it is passing from the Cretaceous to the Jurassic.

* Dino, fierce; Saurus, a lizard. † Enalios, swimming.
‡ River — meuse — saurs. § Pteros, wing.

This system in America holds the remains of the advance guard of the Oysters. The first of the family of Oysters had a shell with curved beak.

In Europe the strata testify that the Jurassic world was astir with reptilian life. Reptiles were in the sea, in the air, and on the land in the groves of Tree-fern and Palm and Cycad.

In the Sierra Nevada, the Uintah, and the Humboldt Mountains there crops from under the Jurassic a still older group of rocks called the Triassic.

Sandstone of the Triassic period lies in narrow belts along the Atlantic coast from the Palisades of the Hudson to Richmond, Virginia; from New Haven to Northern Massachusetts, and from Oxford in North Carolina, over Deep River, into the edge of South Carolina. These belts of sandstone were laid down in valleys, formed, perhaps, when the crust of the earth along the borders of the Atlantic was folded into the Appalachian Mountains. In the Connecticut Valley these Triassic beds are marked by footprints.

On the plains of Babylonia, when Nebuchadnezzar was King, men were moulding clay into bricks, stamping them with the seal of the Great King, and then spreading them in the sun to dry. One day a terrier dog stepped on a new-moulded brick. That brick has been dug up from the ruins of Babylon and we have seen the footprint, which partly effaces the signet-stamp of Nebuchadnezzar. A footprint on a brick is the evidence that terriers were living in Babylon two thousand five hundred years ago.

*Very* long ago, when the site of Babylon was the bed of an ocean and the Alleghany mountains were still young, tides were rolling up through an estuary into the heart of New England, and spreading out over a low, marshy beach, leaves of mud and sand. Living things, in quest, perchance, of what the tide had left, were walking to and fro and stamping their

Fig. 19.

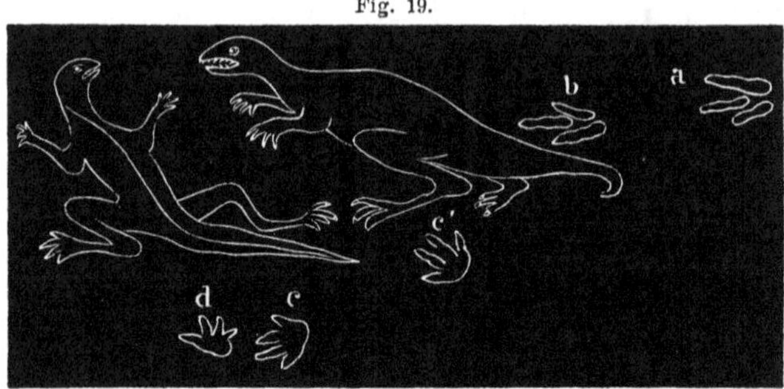


footprints in the mud. That mud has become rock and we have stood on its tracked and upturned ledges at Turner's Falls and tried to call up out of the buried ages a picture of that Triassic life which,

> " Departing, left behind it
> Footprints on the sands of time."

In Fig. 19, at *a*, is represented a footprint nearly two feet long, and at *b* another impression a little more than five feet from the first. The first is a print of the right foot, the second of the left. The animal we are trying to know walked like a Bird, and its limbs were so long it could swing more than ten feet at a stride. If a Bird, it must have stood at least fifteen feet high. But a Bird it was not, only the *fraction* of a Bird. The type of foot which made these impressions belongs to the Bird, but the same type is found in the huge, bipedal Dinosaurs. As we have no evidence

Fig. 19. Scene on the Connecticut Valley in the Triassic Age.

that the Bird had come in Triassic times, complete and separate from the Reptile, we infer that a Bird-reptile with long legs and immense body was walking here when these slates were mud on the beach of an estuary.

Another print is shown in Fig. 19, at *c*. It is that of a four-toed foot twenty inches long. At *d* is seen an impression by the fore-foot, which was small and was not often brought to the ground. The stride of this animal was three feet. Its hind-limb was long and strong, its fore-limb short and weak. The animal

Fig. 20.

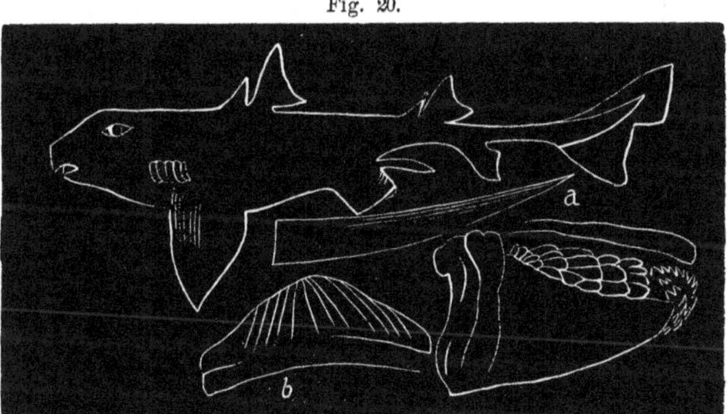


was a tailed bipedal Frog which could stretch itself up to the height of twelve feet or more.

To read the sequence of events in the earlier history of the globe, we must look again at the far West.

In the Elk Mountains of Colorado we find members of a rock system called *Carboniferous* capped by the Triassic. This shows us that the *Carboniferous*, or Coal-making period, preceded the Triassic. Exploring the Carboniferous rocks, we find in them the remains

Fig. 20. PORT JACKSON SHARK.—*a*. Dorsal Spine. *b*. A Tooth.

of a race of Amphibians, the advance-guard of the Reptiles.

Returning to the East, we find under the Carboniferous a group of shale, sandstone, and limestone, called *Devonian*. The same system comes to the surface in Devonshire, England, whence the name.

Exploring the Devonian rocks, we take leave of all forms of vertebrate life except the Fish. Spines occur like that shown in Fig. 20 at *a*, and teeth like that shown at *b*. In many Sharks of the living seas the dorsal fin is armed in front with a spine like this at *a*, and a peculiar Shark off the coast of Australia has the mouth paved with teeth like this at *b*. We infer that spine and tooth belonged to a Shark something like the *Cestraceon* shown in Fig. 20. This Shark of Australian seas, which, like the life of Australia itself, is *old-fashioned*, is only two feet long. The ratio between length of spine and length of fish is as one and a half to twenty-four. If this ratio held from the beginning, our Devonian Shark, since the spine was ten inches long, must have been more than thirteen feet long; and Sharks later down in time, Sharks which dropped their spines into the mud of Carboniferous seas, must have been twenty-seven feet long! It is something to know that the terrible, the all-devouring Shark belongs to a waning dynasty.

Fig. 21.

Fig. 21. The head of one of the early fishes and the whole of one of the last. Outré species came late.

We read in one of our classics that once the Fish was a grass-eater:

> . . . . "And shoals
> Of fish that with their fins and shining scales
> Glide under the green wave, in sculls that oft
> Bank the mid-sea; part single, or with mate,
> Graze the sea-weed, their pastures, or through groves
> Of coral stray."

Science is writing a very different history of creation. From murderous tooth and spine, entombed in the rocks, we read that the primitive Fish was the most terrible representative of the most terrible order of Fishes.

> "Through groves of coral stray"—

the Miltonic imagination struck its wing against a single fact. The early Devonian was a coral-making age. Over vast areas of sea-bottom was a flower-garden of coral-building Polyps. Stand on the ledges of limestone at the Falls of the Ohio and you will recall Florida and what is going on to-day off the Florida coast. Here are coral cups, and domes eight feet in diameter, and clusters of columns, standing where they grew or lying in fragments as they were broken and piled up by the waves.

The reefs that bloomed under primeval seas were not like the reefs that grow now. Not one form of Polyp has come down unchanged from Devonian seas to our own. Many coral-tubes of these ancient rocks are divided by horizontal plates into chambers. *Vertical* plates characterize all recent corals except one group called *Millepores.* Horizontal plates are characteristic of Millepores, which are wrought not by true Polyps but by Medusæ-polyps known under the name of Hydroids. Bell-shaped Medusæ were in the Devonian seas, roving above, while Hydroïds were rooted below.

Other and smaller rock-makers come to our minds out of the Devonian limestones. These limestones take the adjective *corniferous* on account of the seams of hornstone* they contain. Hornstone is flint-like quartz. How was it made? The microscope will tell us. It tells us that this stone is composed of the flinty cases of Diatoms and the spiculæ of Sponges and the teeth of Mollusks.

Under the Devonian is a group of strata called Upper Silurian. It is a record of revolutions achieved through slow movements of the earth's crust. The lowest member of the group (Medina) by its ripple-marks and wave-lines, testifies to having been formed under shallow water; the next member, (Clinton) by its bed of iron-ore tells us of bogs or marshes clad in vegetation; the next member, being shale, (Niagara) was formed on the floor of a deep sea; the next member (Niagara limestone) tells us by its corals that it was built up in reefs on the bottom of a sea, shallow, warm, and placid; the next member (Salina) tells us, by the salt which impregnates it, that the bed of the coral-growing Niagara sea had risen and become a series of lagoons; the highest member of the group (Helderberg) testifies by its corals of another subsidence, and that the lagoons and sea-marshes of the Salina had given place to an open sea.

The Silurian, in America, antedates all vertebrate life. If we can trust negative testimony, the air in that remote epoch was tenantless, the land tenantless,

* Cornu, a horn.

and the seas were tenanted only by Cœlenterate animals, by three-lobed articulate animals which have sent no representative down to us, and by Mollusks, in a wealth of species, of which only two or three have passed down unchanged into our own times and seas.

Fig. 22.

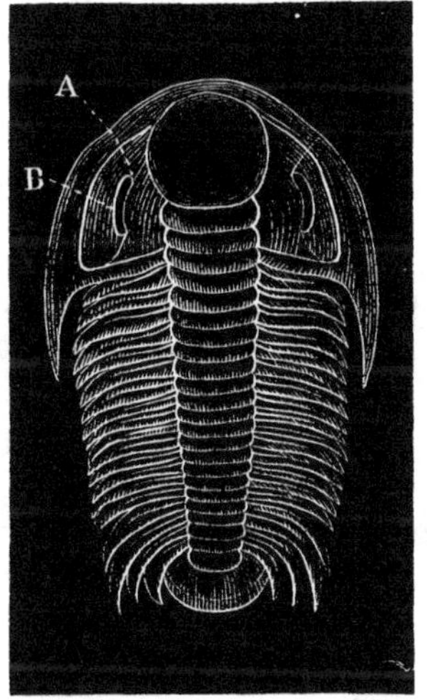


Under the Upper Silurian and separated from it by a change in the dip of the strata and a seeming break in the chain of life, is the Lower Silurian.

In these strata are recorded the events of a past which science can but dimly restore. Of land, and lakes, and rivers, we have lost all trace. We stroll along the shore of a Silurian ocean when we walk over the rippled sandstones of Northern New York and looking *sea-ward* we find the ruins of such poor life as the waters held, but looking *land-ward* we see nothing but rocks made from the ruins of older rocks. We know from the ripple-marks that we are on a shore, but we look in vain for any trace of the land.

In later beds of the Lower Silurian we find shells, in such multitudes as to form almost the entire mass of the rock, and the broken stems of Crinoids and branches of coral. But while the Molluscan type, in its lower orders, had reached its culmination, the type

Fig. 22. Paradoxoides Harlani.

of the coral-making Polyp had not unfolded into the wealth of species which bloomed under later seas.

In the oldest Silurian beds we find hardly a shell that belongs to a living order, and not the trace of a coral Polyp. The world, as we have learned it, is fading out. Plume-like fossils, small and delicate, we find — the ruins of Hydroids born of Medusæ. The Hydroid-medusæ are older rock-makers than Polyps.

Fig. 23.

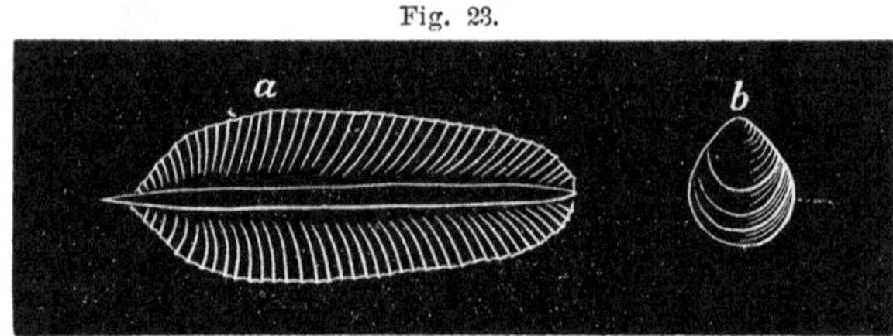


Fossil Sponges we find. The Sponge which hovers on the confines of the animal kingdom and hardly knows itself from a plant, is older than the shell-making Mollusk. Borings and tracks of Worms we find. The segmented worm was on the first sea-shore recorded in the rocks. It was the root, perchance, whence sprang the five branches of the animal kingdom.

Below the lowest bed of the Lower Silurian lies a rock-system of immense thickness which has lately taken the name of *Archæan*. It comes to the surface in the Adirondacks, in the White Mountains, in the Wind River Mountains, the summit ridges of the Rocky Mountains, along the Northeast shore of Lake Huron and along the River St. Lawrence, and is either *on* or *under* the surface everywhere over the globe. It is a universal formation. The rocks are hard and

Fig. 23. Graptolite, *a*, and Lingula, *b*.

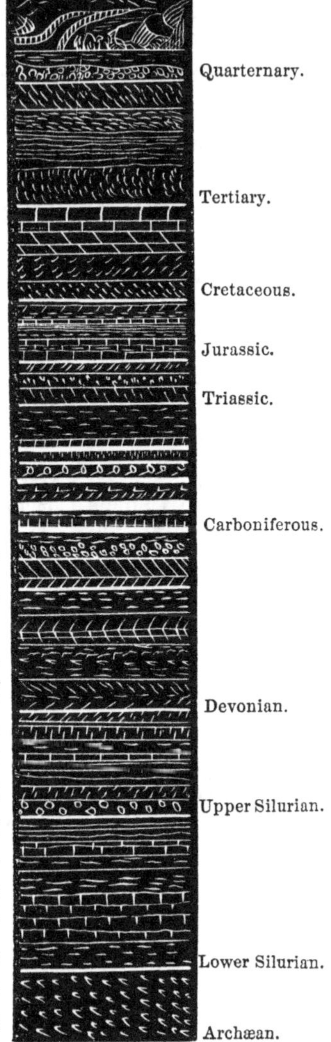

Fig. 24.

crystalline and include granite, and gneiss, and hornblende-slate, and mica slate, and crystalline limestone. In these rocks we find no characters of a sea-bed. The oceans seem to have vanished. The world of life seems to have faded out. What Dante imagined written on the arch over the gate of hell seems to be written on the Archæan rocks.

> "Before *me*, things create were none,
> Save things Eternal."

We have glanced over the rock-sytems as they lie in the earth's crust from the coral reefs of Florida to the Archæan granite of Canada. The book which we have read only by the title page of its chapters is a volume of strata whose thickness has not concerned us. Reviewing them in the same order we find the Alabama marl eighty feet thick; the fresh-water Eocene strata of Wyoming, we find by their exposure along the flanks of mountains and the walls of cañons, are eight thousand feet thick; the Cretaceous system, we find by measuring its upturned edges in Wyoming, Utah, and Colorado, attains a thickness of nine thousand feet; the Jurassic, on the Wind River Mountains, is eight hundred feet thick; the Triassic in the Connecticut Valley, three thousand feet; the Carboniferous, along the Appalachians, sixteen thousand feet; Devonian, fourteen thousand four hundred feet; Upper Silurian, seven thousand three hundred and sixty feet; Lower Silurian, thirty-one thousand two hundred feet.

Adding up the figures we find eighty-nine thousand seven hundred and sixty feet, or just seventeen miles of strata between the coral reefs of Florida and the Archæan granite of Canada. Such an immense thickness of strata is the register of vast epochs of time. We are baffled when we attempt to measure them on a scale of years.

The growth of coral gives us a scale which we can apply with a degree of confidence to all limestone formations. Estimates based on many and careful observations, place the rate of growth of a coral reef at *one sixteenth of an inch a year*. At this rate it would require a thousand years for five feet of upward growth. Now there are coral reefs two thousand feet thick which are alive at the top, and are therefore *modern*. Such reefs cannot be less than *three hundred and eighty-four thousand years old*. As all limestones have passed from the sea through the laboratory of life into a reef, and as we have no reason to suppose that the vital process which builds the reef was more rapid in earlier times than now, we can apply to the past the scale given us by the present.

Very little limestone occurs in the Eocene, or the Tertiary of which it is the earliest member. We will make no account of it now. In the Cretaceous of Alabama we find about a thousand feet of limestone; in the Jurassic of the Rocky Mountains about five hundred feet; in the Triassic, very little; in the Carboniferous of the Alleghanies one hundred and twenty-five feet; in the Devonian of the Alleghanies one hundred feet; in the Upper Silurian six hundred feet; in the Lower Silurian six thousand two hundred feet. Adding up the figures we find eight thousand five hundred feet of limestones. Applying the scale, it would give, for the growth of these limestones, one million six hundred and thirty-two thousand years. These figures are very low. They express only the *minimum* of time required for the formation of eight thousand five hundred feet of strata out of a total thickness of eighty-nine thousand seven hundred and sixty feet.

But we have not reached the beginning. The Archæan rocks are *not* inscribed as the portals to Dante's Inferno.

One of the life-forms which first engaged our attention was the shelled Rhizopod, or Ray-streamer.

We have now to project our minds into the abyss of a past eternity and look into the oceans of the Archæan world. We see a mass of animated jelly, without parts, or organs, or senses, absorbing lime and by unconscious powers of masonry building it into a globular shell, perforated for the protrusion of root-like feet. A portion of the plastic jelly exudes through the pores and forms a flocculent mass around the shell. So far, it has behaved like a Globigerina. The animal has simply overflowed its house. But now the outside portion which envelops the shell proceeds to house itself by secreting from the water another shell which follows in the main the contour of the inner one and like it is perforated. The enclosed jelly continues to grow and again exudes through the pores and envelops the shell. The rind of enveloping jelly secretes another shell and this process of shell-making and overflowing goes on until the Rhizopod becomes a mass of shell and jelly two or three feet in diameter. The Rhizopod dies and particles of serpentine infiltrate through the pores and fill up the spaces occupied by the body. Ages pass and the sediment in which the dead Rhizopod lies is buried deep under other sediments — so deep that in the heat generated by their pressure it is changed into crystalline granite. Ages pass, continents rise, mountains are lifted up, valleys eroded, life unfolds from type to type till at the last it culminates in Man. Himself the last, he takes from the Archæan rocks of Canada the *first*, the "Eozoon," the Life-dawn.

Fig. 25.

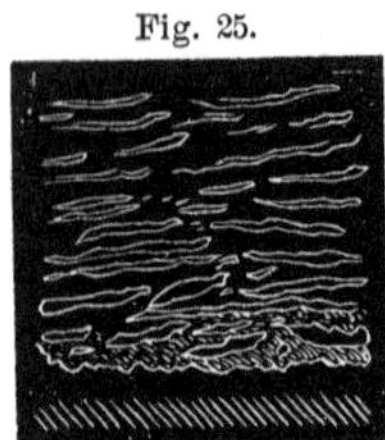

Fig. 25. Eozoon. — The white lines represent the shell, the dark spaces the sarcode.

## CHAPTER III.

The Rotting of Rocks—Time-scale applied to the Archæan Rocks—The Dawn of Life—Evolution of the Cœlanterate Animal, of the Molluscan, of the Articulate, of the Vertebrate—Pedigree of the Fish—History of Petroleum—The Dawn of Plant-life—Evolution of the Lichen, the Scale-moss, and the Fern—The Primeval Forest—The Coal-making Age—Chemistries of the Sunbeam

IN THE air of the Archæan world were agents of destruction not now in the air. Our atmosphere rots organic matter. The primeval atmosphere rotted inorganic matter. It held vast stores of carbonic acid.

The coal locked up in the rocks of North America would form a cubic block twenty miles or more in dimensions. Seventy-eight per cent. of this is carbon. An equal store of carbon, Dawson estimates, is diffused through the Archæan rocks. Once all this carbon was in the air, combined with oxygen and forming a gas known as carbonic acid. Recent investigations place the amount of this gas in our atmosphere as low as three parts, by volume, to ten thousand parts of air. Now acid attacks alkali and even so weak an acid as carbonic assails and dissolves out the alkaline constituents of granite. These constituents are found in the feldspar, which is silica combined with alumina and soda or potash or lime. Feldspathic rocks of the Archæan age, which lay under an atmosphere loaded with carbonic acid, must have yielded their alkali to the acid and crumbled into dust. The rot of such rocks, washed into the ocean, "suffered a sea-change" into other rocks not so rich in feldspar. The granitic rocks, as Hunt has shown, are characterized from the oldest beds upward by less and less feldspar. As we determine the relative age of fossil-bearing strata by their fossils, so we are in the way of fixing the time relations of members of the Archæan system by their minerals. For we must assume that in times the most remote the world was under chemic laws not different from those which at like temperature and with like elements would rule to-day.

And if we assert continuity in the physical history of the globe, we do violence to our reason if we assert *discontinuity* in its organic history. If the chain has not been broken, if new laws have not from time to time been imposed from without, why should we look for breaks in the chain of life and the introduction from time to time of new species from without? The Life-dawn animal entombed in Archæan rocks was related to the life that now is, as the sea on whose bed it lay, and the chemic forces acting through sea and sky and rock, were related to the seas and chemistries that are. Eozoon was like the jelly-speck that forms the Globigerina-shell which the dredge brings from the bed of the Atlantic, and differed from it as a mass differs from a speck, and as a stream that overflows its banks, from a rill that never swells beyond its containing bed.

From the Rhizopod the distance is not great, in one way, to the simplest Worms without blood, and in another way to the bloodless Polyp-jellies. And now the oldest fossils we find, after Eozoon, are plume-like impressions of Hydroids, Worm-tracks and burrows, and Lamp-shells. The Lamp-shells — Lingulæ and Terebratulæ — which linger in a few species

Fig. 26.

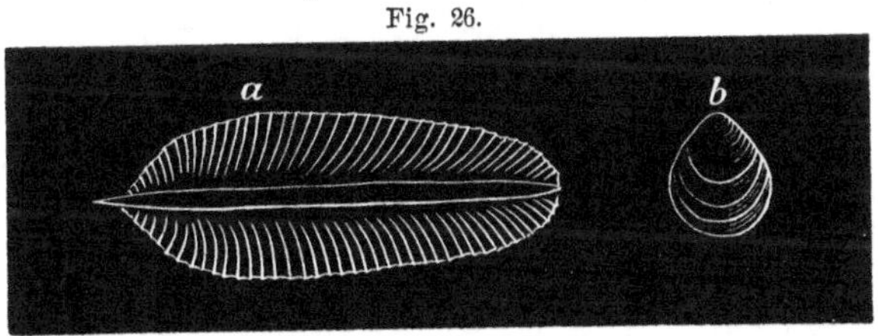


attached to the bottom of the Atlantic, are shown by recent studies to be shells of Worms, not Mollusks. In the Old Silurian seas such Worms were prodigious in number of individuals and variety of forms and their symmetrical bivalve shells, packed together, make up here and there a great part of the Silurian rocks. The gills were spiral and in some species could be extruded, while the gills of bivalve Mollusks, common in our seas and rivers, are leaf-like.

Another primeval type which unfolded into many species and culminated in Silurian seas, was the Crinoid. And the Crinoid, like the Lamp-shell, has survived through all the revolutions of the globe and lingers in little garden-beds, here and there, on the floor of deep oceans to-day. The Crinoid can be understood only through studies on the Star-fish.

In the slates of Hürzbach, the base of the Silurian, fossil Worms have been found which were mailed in calcareous plates like those on the arm of a Star-fish. Now the arm of a Star-fish, when detached from the body, leads an independent life and appears like a

Fig. 26. *a*, Graptolithus Logani; *b*, Lingulella prima

mailed worm. Is a Star-fish, then, a cormus of Worms? We cannot do better here than to put ourselves under the guidance of Haeckel, and accept his answer—greatly abridged.

In a certain class of Worms a number of individuals unite and grow together at their posterior ends. Communism commences at the tail. Each Worm retains its own mouth, but the united members form a common outlet for discharges. Now if these worms were clad in calcareous coats of mail, and were still further *de-individualized*, having lost each its own mouth, and if they were arranged symmetrically around the central body-disk, they would form a Star-fish. From the egg of a Star-fish comes the young Star, not at first a star but a ciliated animal like the larva of a Ring-worm. The first stage in the life-history of an individual is a reminiscence of the remote ancestral form. The Star-fish, we infer, is a cormus of articulated Worms, and has originated by the star-wise growth of buds from a common Mother-worm.

Crinoids were Star-fishes which, becoming sedentary, developed stems and attached themselves to the sea-bottom. Their stems were made of little calcareous disks perforated through the center and placed disk on disk. They clasped the sea-floor by root-like holdfasts which are often found in the limestones of Crawfordsville, Indiana. Crowning this stem was a body composed of many close-fitting plates and taking many shapes, generally that of a cup, with branching rays fringed with pinnæ, radiating from the rim. When it spread its rays the Crinoid looked like a flower. In sheltered nooks on the coral floor of primeval seas, they spread their rays on the water, generation after generation, age after age, until their fallen stems and cups had formed immense beds of limestone. The builder little knows what builders were before him.

Fig. 27.

Standing in the world's great cathedral we have wondered whether.

> "The hand that rounded Peter's dome
> And groined the aisles of Christian Rome,"

had more cunning than the unconscious builders of the

Fig. 27. Platycrinus Safordi.

rocks which it has built into temples. For the great cathedral was coral-made and crinoid-made before it was man-made.

A member of the articulate class, removed very far from the Worm, appears fossil in the oldest beds of the Silurian. A standing challenge to science — whence it came and why it perished — is the Trilobite. Trilobites were three-lobed Crustaceans with thin, leaf-like legs — if any legs at all — and compound eyes. Very few members of the family crossed the confines of the Devonian, and not one crossed the Carboniferous. The earliest recorded Trilobites, as those found in the slates of Braintree, were the largest of the family. Whence came they? We must project our vision beyond the times recorded in the Potsdam sandstone into the Archæan age. If we could restore the first record on that palimpsest, the Archæan system, we can hardly doubt the remains of Trilobites would be there, and the ancestors of Trilobites.*

Fig. 28.

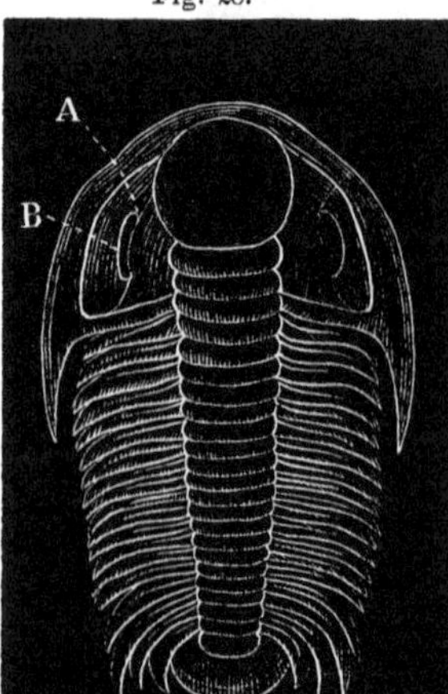


Rhizopods, Sponges, Worms, Polyps, Mollusks, and Trilobites — with such poor dower of life the world

* Palimpsest: "A manuscript which has been written upon twice, the first writing having been erased to make place for the second." The fossil record, save Eozoon, has been effaced, and this system is a vast record of metamorphisms.

Fig. 28. Paradoxides Harlani. *A*, Movable cheek. *B*, the Eye.

got on through the ages of the Silurian to the Devonian. No brain-bearing head did the world have till the advent of the Devonian Fish. How long was there life without brains?

The Archæan rocks in Central Europe have a thickness of ninety thousand feet — more than nineteen miles — and Eozoon has been found at their very base! Time estimates in Geology can be only crude approximations to truth. From all the data at command some authorities have estimated twenty million years for the growth of a mile of sediment over a sea-bed. At this rate we must allow three hundred and eighty million years to the Archæan age, and forty millions more to the Silurian. Three hundred and eighty million years of life without consciousness! Forty millions more of life without brains! Four hundred and twenty million years — did the globe, having got life, wait four hundred and twenty million years for *brains?* Did it take creation four hundred and twenty million years to get under *headway?*

These numbers are appalling. If the mind recoils from them let it attempt to picture the work against which they are laid as a time-scale: Twenty-one miles of sediments, and (as sediment growth under the sea implies rock destruction under the air,) rock-beds to the thickness of twenty-one miles rotted and abraded into dust — and as wide as the area of subaqueous sediments, so wide the area of subaerial destruction. The problem is complicated. The nineteen miles of Archæan rocks in Europe can not represent a continuous destruction of rocks on a continent, or a continuous growth of sediment under a sea. And then metamorphism — what titanic forces wrought on these sediments to transform them into crystalline granites, and what is *their* time-scale?

Fig. 29.

Fig 29. Cephalaspis Lyelli.

The earliest recorded Fish was no more the first of its class than the Braintree Paradoxides was the first of Trilobites. A buckler-shaped plate studded with bony stars shielded the head, and rhombic plates, placed like tiles, shielded the body. The eyes were placed far up on the head and close together. The form of this Fish will be seen in Fig. 29.

Gegenbaur has lately shown that in the structure of head and limb, heart and intestine, the order to which the oldest of recorded Fishes belongs is related to the order of Sharks. Cranium and fin and heart and intestine are related to the corresponding parts of a Shark as developments or reductions. The enameled Fish, or Ganoid, implies, then, some form of Shark as its precursor.

In brain and gills the Shark is related to the order represented by the Hag and Lamprey, an order characterized by absence of fins and scales and swim-bladder and jaw-skeleton. The precursor of the Shark, we must infer, was a Fish organized on the plan of the limbless, scaleless, jaw-less Lamprey.

Animals of the Lamprey's order — they are hardly Fishes — represent a stage between the true Fish and the Lancelet. As Fishes assume in embryo the form which is retained through life by the Lancelet, we place, farther back in time than the farthest Fish, a lance-shaped animal with an unsegmented cartilaginous sheath — the first draft in all vertebrates of a spinal column. No record of such an animal, and no record of the Lamprey-like half-fish which came after it, has been inscribed on the rocks.

All primeval Fishes had cartilaginous skeletons. All recent Fishes have bony skeletons. In bony

Fishes the departure from the Shark taken by the Ganoid is carried still further. The precursor of the bony Fish, we must infer, then, was a Ganoid.

Hints of the way in which creation traveled along this path are not wanting.

A station on the way from Shark to Ganoid was a Fish which we represent in Fig. 30. The tuberculated

Fig. 30.

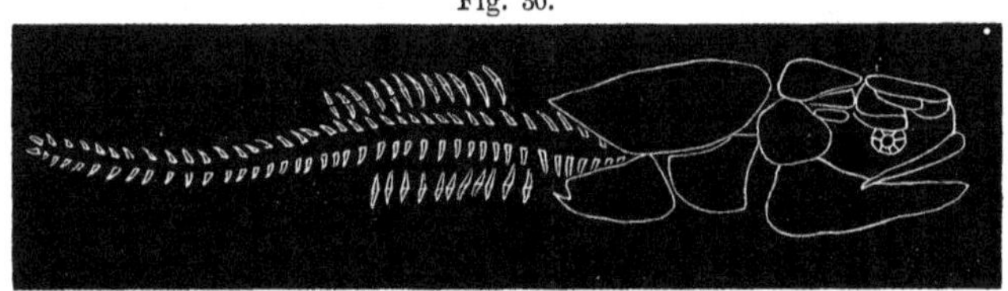

plates which partly covered the body, the ribs, and the arches of bone which clasped the marrow tube (notochord) are preserved in the rocks and shown in the plate. Between the ribs and neural arches is a blank, which shows that the center of the spinal column was cartilaginous. Already part of the internal skeleton was bony.

From a Fish with vertebral column on the same pattern as this, Huxley has pointed out some of the steps toward a modern Fish with the vertebral column bony in the center as well as spines. Arches and ribs were distinct from each other as they were in the Fish shown in Fig. 30, and were united to the notochord by ligaments. In species of the same family found in later strata the arches and ribs were expanded at their ends, showing a tendency to embrace the notochord. In still later species the process of flattening was carried so far as almost to produce a bony ring like the rudimental body of a vertebra.

Fig. 30. Coccosteus decipiens.

Ancient Fishes had vertebrate tails. Generally the column extended through the upper lobe of the tail fin. Modern Fishes have invertebrate tails. The column ends where the tail fin begins. In Triassic strata we find the remains of Fishes with tails half vertebrated, and remains of others with tails not vertebrated at all. In a deeper sense than Agassiz meant when he wrote it, "the progress of the ages is marked in the tails of Fishes."

The Corniferous limestone which, in America, holds the oldest fish-remains, and which, as shown at the Falls of the Ohio, was a vast coral reef that spread over the floor of the Devonian sea, and, as shown everywhere by its hornstone, received the flinty cases of Diatoms that fell from above and the spiculæ of Sponges that were rooted to the coral below, is rich in a treasure of whose origin its own life-origin may bear testimony.

Rock-oil, even in its relations to man, is no new thing under the sun. More than two hundred years ago it was known in Italy and used to light the houses of Parma and Modena. Taken from the wells of Rangoon it was used as a medicine in Burmah more than two thousand years ago. It flows up in a chain of springs (slime-pits of Bible,) along the margin of the Dead Sea, and mingling with its waters makes them heavy and bitter and vile. It oozes out from the base of the volcanic islands of Cape Verd and bubbles up through the sea off the northern slope of Vesuvius. It saturates the soil of a little peninsula that juts from Persia into the Caspian, and, oozing from the side of a mountain, is collected in reservoirs hewn in the rock in some unknown antiquity. At the base of the mountain it leaps up in a pyramid of flame twenty feet high. "The eternal fire of Bakou"— there it burns now and there, perchance, it burned in the days of Zoroaster. There, perchance, the early Persian looked and wondered, and his wonder being exalted into worship, flame became the symbol of Deity. The Fire-worshipers built their altars over springs of burning gas like this, a fact which would lead us to infer that the wonder excited by such spontaneous burning led to the worship of fire.

In Ecbatana Alexander saw a gas flame issuing from the earth, and wondered at the sight. Plutarch's account seems to bring these old times forward into our own. "Alexander was much surprised at the sight of a place where fire issues in a continuous stream from a cleft in the earth, and where a stream of naphtha not far from the same spot flows out so abundantly as to form a sort of lake. This naphtha, in other respects resembling bitumen, is so subject to take fire that before it touched the flame it kindled at the very light that surrounded it"—Truthful Plutarch! This statement attests his veracity. He knew nothing of the gas escaping from and enveloping the naphtha, which he describes by its effects. He has won our confidence and we will let him go on—"and inflamed the intermediate air. The barbarians to show the power and nature of it sprinkled the street which led to the king's lodgings with little drops of it and when it was almost night stood at the further end and with torches lighted the drops, and at once, as quick as a man could think, from one end to the other it took fire in such a manner that the whole street was one continued flame. An Athenian who attended

Alexander in his baths desired to make an experiment of the naphtha upon Stephanus, a young man with a ridiculously ugly face, whose talent was singing well. 'For,' said he, 'if it take hold on him and is not immediately put out we must allow it to be of the most inconceivable strength.'" Plutarch has our confidence still. This reads like an incident in modern camp life. Let him go on. "The youth, as it happened, readily consented to undergo the trial and as soon as he was anointed and rubbed with the naphtha" — hold! This will hardly do. That any young man "with a talent for singing" should be such an incomparable fool as to submit to such an experiment — we would not believe it if Plutarch had not told us of the naphtha's kindling "at the very light which surrounded it" and "inflaming the intermediate air." On the faith begotten of that statement we will hear him through. — "As soon as he was anointed and rubbed with the naphtha he was touched off and his whole body broke out into such a flame that nothing could have prevented his being consumed by it if, by good chance, there had not been people at hand with a great many vessels of water.

"The manner of the production of this naphtha admits of diverse opinions"— Here there is a blank. Some of Plutarch's words have been lost. It was so good in one of the old monks to lose the manuscript at this very clause. For what could the good man have told us about "the manner of production?" Or what do we care for "the diverse opinions" of his neighbors on a matter which does not admit of divers opinions?

The burning gas of Bakou and Ecbatana was an exhalation from a tomb. If Plutarch had known this essential fact the missing words of his manuscript might have told us the origin of the carbon gas. The rocks which hold the hydro-carbons in America are the corals and shells and stems of organisms vastly older than those which formed the oil-giving rocks of Asia.

It may not be enough to know in a general way that the hydro-carbons, gaseous and fluid, come from strata made of fossils. Let us look at details. At Watertown, New York, we find petroleum trickling from the pores of coral in Trenton limestone. In the same limestone, cropping in Canada, we find it filling the chambers of orthocerata. On the shores of Lake Erie we find it, like honey, filling the pores of a honeycomb-like coral. In Western New York we find it distilling from the immense stores of carbon diffused through a member of the Devonian system called the "Genesee shale." Its grosser part is carbon, its etherial part is hydrogen. Here and there, over the region underlaid by Genesee shale, the hydrogen, bearing up atoms of carbon, rises into the air as a gas-spring like that which surprised the Macedonian king in Ecbatana. Now carbon in all its visible and tangible forms is a product of life. Hydrogen, as all the world has learned, is a constituent of water. The hydrogen of carbon oil may have come by distillation from the sea. From what forms of life came the carbon? As the oil occurs in strata such as the Utica slate, which seem to antedate all land life, it could not have been derived from land plants. Sea plants may have yielded a portion of it, but as the occurrence of the oil in fossil shells and corals testifies, it could not have yielded it all. Lower animals contain carbonaceous tissue analogous to that of plants, and, as was first shown by Hunt, the remains of such animals, by slow decomposition, may give rise to the hydro-carbons.

As the chief petroleum-bearing strata are composed of the shells and corals and stems and cups and flinty spiculæ of the low and early forms of life, the oil, which represents the tissues of the animal, should be in some sense commensurate with the strata, which represent their skeletons. The product of the Pennsylvania oil region between 1860 and 1870 is estimated at twenty-eight millions of barrels. Dr. Hunt has analyzed specimens of the oil-bearing limestone under Chicago, and shown that although this rock does not yield oil in quantities that pay the working, each square mile of it contains seven and three-quarter millions of barrels.

Petroleum is a gift of the sea. In rocks higher up on the geologic scale than those which, in America, yield the oil, we find a product allied to it — the first gift of the land.

The rocks which were first to register the Fish, register the first appearance of land-plants.

We have pictured the primeval Fish from the Shark. We must picture the primeval land-plant from the little Club-moss whose branching stalk curves and creeps on the ground, and sends up pedicels which bear the spores in cylindrical spikes. Closely related to the Club-moss (and all Mosses) is the Fern, and close on the heels of the Moss came the Fern. Frondose-ferns (Tree-ferns) have been found in the Corniferous limestones of Ohio.

Were Club-mosses and Ferns first in the race of land-plants? One bond of kinship between Fern and Moss is the fore-growth. Every Fern and every Moss represents two generations. In the first, stem, root, and leaf appear as one member, called a Thallus. This is the fore-growth. From the fore-growth comes the after-growth, differentiated into stem and root and frond. In the lowest Mosses these parts remain still involved in the thallus. Every Fern and every higher Moss, in its beginning or fore-growth, is a kind of Liverwort, and we infer that the class of Thallus-plants must have preceded that of the Ferns and Scale-mosses—although the rocks do not testify. The lowest of Thallus-plants is the Lichen and recent investigations have shown that Lichens are composed of two distinct plants, Fungi and Algæ. The Fungi are a large class, reaching from Mushrooms and Toad-stools to minute specks which form mould and mildew and smut, and cause fermentations. They are all parasites, and live not as other plants, on inorganic, but as animals, on *organic* matter. Algæ are cellular water plants, leafless and flowerless. They are in the sea in many forms, some of surpassing beauty, and, as told by their prints on the strata, they were in the Old Silurian sea, and, as told by the carbon in Laurentian rocks, they were in the Archæan sea. Now a Lichen is composed of a low form of Alga and a Fungus which lives on it as a parasite. Lichens then presuppose Algæ, and Algæ were born of the water. The gift of life came to the land out of the sea. Let a volcanic island rise from the sea and the lava grow cold and hard. Seeds and spores, borne on the wind, may be dropped on it, but they cannot germinate. Its flinty face will resist every form of vegetation but the Lichen. The encrusting Lichen comes and, adding cell to cell, presses against the rock, detaches from it atoms for a soil, and wins something from the realm of death for the uses of life. Its decayed thallus, with detached atoms of rock, forms the first mould for the growth of Mosses and Ferns. So the Lichen, a fringe of gray over all the nakedness of nature, first life on new-risen land to-day, was first on the first land that emerged from the sea.

The plant kingdom, established on the land earlier than the Devonian age, unfolded and reached its culmination in the flowerless orders during the age known as Carboniferous.

There were movements in the earth's crust, slow but vast, which changed the relations between sea and land, and modified the life of both. Beds of Devonian strata are found tilted up, and in places metamorphosed into crystalline gneiss. Overlying such beds, unconformably, are beds of Carboniferous age. The disturbances through the earth's crust closed the Devonian age and ushered in the Carboniferous.

Uneasy times preceded, uneasy times followed. Vast oscillations of the earth's crust characterized the Carboniferous epoch. Downsinkings and uprisings over large areas, and repeated many times, are registered in Carboniferous strata. Where these strata have the fullest development they testify of two great movements to which all other movements were subordinate. The typical Carboniferous system consists of, first and lowest, beds of limestone; second, conglomerate and gritty sandstone; third, the coal-measures. In England and Germany the coal-measures are followed by, first, conglomerate and gritty sandstone; and second, beds of limestone. The limestone is made largely of Crinoidal remains and testifies to the culmination of this beautiful order. The limestone was formed on the bed of a clear and open sea. The conglomerates and gritty sandstone (Millstone grit) were shore and beach deposits. The coal-beds were the growth of marshes. Here is a record as legible as any record written by man. It tells of the elevation of the bed of a Crinoidal sea, until, as it approached the surface, it was swept by currents and gritty sand was spread out over the emerging wrecks of Crinoids, and pebbles were piled up along the shores. It tells that the upward movement was arrested before the sea-bed became dry land, and that the surface remained stationary while the marsh vegetation of the coal-beds was growing. It tells us that after this long rest a downward movement began, which carried the accumulations of marsh growth below the sea-level, where, swept by currents, it was overspread with sand like that which the emerging sea-bed had received ages before, and that the sinking of the crust went on until the site of the marsh had become the floor of a quiet, lime-making sea in which the Crinoid was now a waning dynasty. This is the cycle of the Carboniferous and "Dyas." The strata in which only part of this story is told attain a thickness, in Nova Scotia, of fourteen thousand five hundred feet! Over large areas these movements were fitful and interrupted. Where Nova Scotia is there were seventy-six oscillations of level.

The transition from sea to marsh is written in the Millstone grit. Each coal-bed is a register of the time through which the land surface remained a marsh.

Over vast marshes or "dismal swamps" spread a vegetation heralded by the Club-moss and Fern of the

Fig. 31. America in the Carboniferous Age.

Devonian. The type of our creeping, much-branching Club-moss had culminated in trees whose columnar trunks, embossed with scars and branching at the top, rose sixty or eighty feet in the air. The Scale-tree threw out its branches in pairs and each branch forked and the branchlets forked again and were crowned at the summit with tufts of scaly leaves and, like the trunk, were embossed with spiral lines of scars which marked the place of fallen scales.

By the side of the Scale-tree rose the Frondose Ferns, beautiful to-day in the steaming forests of the torrid zone, but richer in Carboniferous days in the tracery of their fronds, and more stately in their pillared stems, as they stood over every steaming bog in every zone from Greenland to the line.

By Frondose Fern and Scale-tree and overtopping each, in likeness of a Corinthian column, stood the Seal-tree, its flutings bearing scars like the impress of a seal. The column spread atop into two branches crowned with fronds like a Frondose Fern, and was braced by a wreath of subaqueous stems. The tree, composed of two orders (Club-moss and Tree Fern) has taken two names, *Sigillaria* for the above-water stem and *Stigmaria* for the under-water stems.

In stems, jointed as well as fluted, and rising to the height of fifty feet, were the primeval Reeds.

On drier ground stretching away from the marsh, in the form of a stunted Palm, grew the Cycas, prophetic of Palms, and Conifers prophetic of Pines.

To the land forest of Coniferas and Cycads and the marsh forest of Scale-trees and Seal-trees and Reed-trees and Fern-trees, add an undergrowth of low herbaceous Ferns and you have the picture of a primeval landscape. Blot from the face of nature every flowering weed and every flowering tree, every grass, every fruit, every growth useful to man or beast; go then to the Sunda Islands for the largest Club-moss, to the East Indies for the largest Tree Fern, to the damp glades of Caraccas for the tallest reeds, to the Moluccas for their Cycad and to Australia for its pine, to the ponds and sluggish streams of Europe or America for their Quill-wort, and place them side by side over a vast marsh and its sandy borders and you will faintly *realize* your picture of a primeval landscape. Dwarf the Cycad and the Pine, lift still higher the tapering column of the Tree Fern, multiply by two the bulk of the Reed and by three the bulk of the Club-moss, lift the Quill-wort from the water and to its long linear leaves add a fluted stem eighty feet high (for the Seal-tree) and you would *fully* realize a Carboniferous landscape — realize in all but its vast solitudes. Not a bird ever perched on spiky leaf or spreading frond of a coal forest. No flower had opened yet to shed fragrance on the air and no throat had warbled into its stillness a note of music. Such poor animal life as possessed the Carboniferous world has left its imprint on wave-washed shores and in the hollow stems of fallen trees.

Seal-trees have broken above their water-stems and fallen into the bog. Their hollow stumps have become a hiding place or hunting ground of Amphibians whose bones and body-scales and sculptured crania have been taken, fossils in a fossil stump, by Dawson and Lyell.

Walking on a shore or hiding in a stump — no matter — the earth will record the one as well as the other and hold the record forever! Fig. 32 represents the surface of a stratum in the Carboniferous system of Pennsylvania. Something walked there when the rock stratum was soft mud. Its fore-foot was four inches broad and shaped something like a hand with five stubby, spreading fingers. Its hind-foot was four toed and in walking was placed nearly in the track left by the fore-foot. The stride was thirteen inches. From the character of the feet we know that this ani-

Fig. 32.

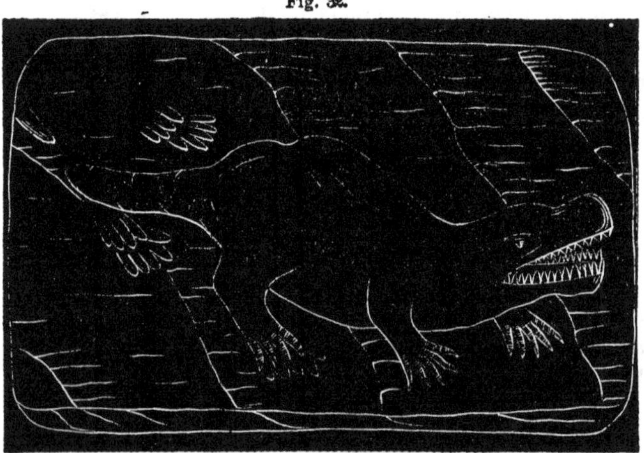

mal was an Amphibian belonging to an order which has become known to us through fossil crania and teeth, as well as foot-prints. The animal was a *Labyrinthodont*, and although itself not remarkable for size some of its kindred attained the bulk of an ox. It walked here on a beach invaded now and then by the waves. Wave-marks appear on the stratum which were partially obliterated by the tread. We infer that

Fig. 32. A Labyrinthodont and its tracks on a ripple-marked beach.

waves had swept over the flat not long before the Amphibian walked. The surface of the stratum, even on the ripple-marks, and in the foot-prints is dotted over with the impressions of rain-drops. Waves had rolled up against a beach and swept over a bordering flat. An animal to which the living world has nothing near of kin, followed the waves and, here and there, beat down with its foot-prints their piled-up ridges of sand. A rain-storm followed the animal, and the drops, pattering on the rippled and tracked mud, left *their* prints — the last event in that record of a Carboniferous day.

Tenanted by the earliest patterns of Centipede and Scorpion and Spider and Cockroach and May-fly, and tenanted by Labyrinthodont Amphibians, the forest of Ferns and fluted Scale-trees and Seal-trees endured through centuries whose only register is the coal-bed formed from its ruins. As the Sphagnum growing in a bog, dying, does not decay but accumulates, generation after generation, in the form of peat, so these marsh forests of the Carboniferous age, dying, still endured.

As it requires between five and ten feet of vegetable debris to form one foot of coal, we must imagine the forest registered in a coal-seam to have endured long enough for the formation of debris five or eight times the thickness of the coal. The Pittsburg seam is ten feet thick, and a seam under Pictou, Nova Scotia, is thirty-five feet thick. The one represents accumulations of vegetable debris not less than fifty or more than a hundred feet in thickness, and the other, accumulations between a hundred and seventy-five and three hundred and fifty feet in thickness. We have no data from which we can estimate the time required for such growth. Such time would represent a stand-still period. Such a period was followed by one of submergence. The surface, loaded with the growth of the quiet centuries, was carried down beneath the sea where it was swept by waves and overspread with sand and mud, or carried deeper where it was overrun with lime-making Polyps and Crinoids. To a period of submergence succeeded one of emergence. The bed of the sea rose again, and again became a marsh, the site of another Fern forest. Scale-trees spring up again, but not in such bulk or profusion as those which had grown there ages before, and whose ruins were locked up now in a seam of coal below.* Ferns, herbaceous and frondose, came again over the deep-buried remains of their ancestors. Seal-trees returned, and Reeds, recruited by new forms which bore their leaves in stellate whorls from the joints or nodes.

And so the ages passed, ages of rest, until vast stores of carbon had been taken again from the air and stored as vegetable debris over the marsh. Another submergence carried the land again below the sea, and to another era of verdure succeeded another era of desolation. Beds of sandstone or shale or limestone, overlying a second seam of coal, record the second period of submergence.

For the third time the bed of the sea emerges and

* The spores of Lepidodendrons, something like the spores of our Lycopodium used in fireworks, are so abundant in the coal as to have led certain observers to think that the mass of the coal was Club-moss spores!

stands for the third time at the water-limit, not now a region of verdure but a region of barren marshes in which, by a process described in our first chapter, bands of iron-ore were formed. Every coal-seam represents ages of accumulation over verdant marshes. Every bed of iron ore represents accumulations in barren marshes. Every intervening bed of limestone, or shale, or sandstone, represents an intervening period of submergence. In the Carboniferous rocks of Nova Scotia, better than anywhere else, we can see the scale of these oscillations.

Following the beach at the base of a line of cliffs between one and two hundred feet high, we pass along a series of strata represented, in part, by Fig. 33. We follow the direction of dip, which is southward. At

Fig. 33.

*a* we find beds of sandstone, limestone, and gypsum, dipping southward at an angle of twenty-seven degrees. Following the dip, we pass for three miles along the base of a cliff composed of red sandstone and marl, represented at *b*. Following on in the same direction, we pass along upturned beds of sandstone and shale and coal and limestone—*c*, *d*, *e*, *f*, *g*, *h*—stratum after stratum, in all seven thousand feet. Within these seven thousand feet of shale and sandstone and lime, we find seventy-six coal-seams, and under each coal-seam, a layer of dark clay,

Fig. 33. Section of Carboniferous strata exposed along the coast of Nova Scotia.

called a "dirt bed." It is the fossil soil on which the Fern forest grew whose ruins are preserved in the coal above. We cannot be mistaken in this, for we find the stumps of Seal-trees with the roots penetrating the "dirt bed." It was in one of these stumps that Lyell and Dawson found the fossil Amphibians. In every one of the seventy-six dirt-beds we find fossil stumps and roots, but over fifteen of these beds we find only a trace of coal. The stumps are tilted with the strata.

This record needs no interpretation. It cannot be misread. The six thousand feet of strata we passed over from *a* to *c*, beginning with limestone and ending with gritty sandstone, were formed under an ocean, and the gritty sandstone records the emergence of the beds of that sea into a marsh whose soil is now the dirt bed *d*, and whose ages of forest growth are now the coal-seam *e*. The Sandstone *f* records a submergence, and the dirt bed *g* another emergence, another marsh, and another forest of Scale-trees, but now, as there is only a trace of coal above, an abortive attempt at coal-making. The shale *h* records — but we are interpreting what interprets itself. Seventy-six oscillations between marsh and sea or lake, and these movements extending through a period long enough for the formation of seven thousand feet of strata, and then, by some titanic force, all these strata, with all that lie below, tilted up at an angle of twenty-seven degrees! — this is the account the Nova Scotia coal-system gives of itself. The shrinking and crumpling of the crust in which these beds were tilted at such an angle were as nothing to the movements recorded over the region of the Alleghanies. The coal-beds in the anthracite region of Pennsylvania, and their containing strata, are flexed and doubled back on themselves. We give a section, drawn by Taylor, representing the flexures near Nesquehoning. Some of the strata, it will be seen, are nearly vertical, and the great coal-bed 1 is flexed and folded back on itself. The bed 2, which

Fig. 34.

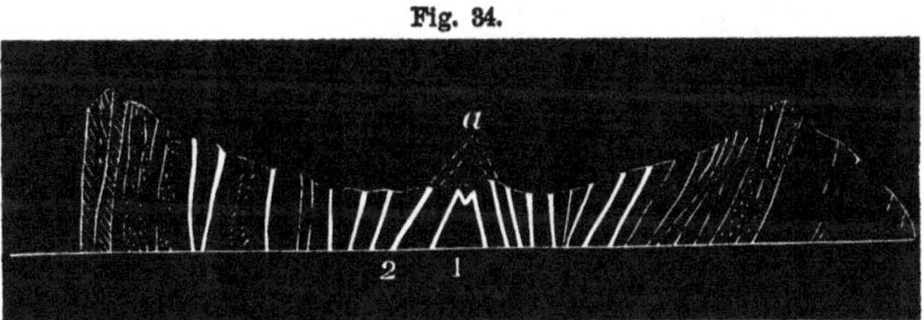


has the same dip as 1, reaches the surface and is cut off but appears again with opposite dip on the other side of the axis. We draw a dotted line along the angles of dip, connecting these seams as once they *were* connected. Our dotted lines would meet at *a*, and from this point to the ground the coal and containing rocks have been stripped off by erosion.

In such crumpling the crust was often fractured and immense displacements and erosions have followed. Near Chambersburg, Pennsylvania, a fault occurs which has been described by Lesley: "The western side of the anticlinal cove-canoe has been cut off and carried down at least twenty thousand feet into the abyss, along a fracture twenty miles in length; the eastern side must have stood high enough in the air to make a Hindoo Koosh; and all the materials must have been swept into the Atlantic by the denuding flood. The evidence of this is of the simplest order

Fig. 34. Section representing the folding of strata in Nesquehoning, Pa.

and patent to every eye. Portions of the Upper Devonian wall against the lower portions of the Lower Silurian. The thickness of the rocks between is, of course, the exact measure of the downthrow, which is therefore twenty times as great as the celebrated Pennine Fault in England. Yet a man can stand astride across the crevice, with one foot on the Trenton limestone and the other on the Hamilton slates, and put his hand on some great fragments of Shawangunk grit, caught as they were falling down the chasm, held fast in its jaws as it closed, and revealed by the merest accident of lying suspended in the crack just where the plane of denudation happened to cut it."

In the flexings and faultings which attended the revolutionary period which closed the Carboniferous age, the coal beds were changed more or less in composition and texture. Where the enclosing strata have not been flexed the coal is bituminous. Where the strata have suffered plication the coal has been more or less debituminized. Where the strata have suffered metamorphism, as in Rhode Island, the contained coal has been debituminized and metamorphosed into graphite.

The far-off age of coal-making was as unlike to ours in climate as in vegetation and conditions for the storing up of vegetation. From the Bolivian Andes near the line to Kotzebue Sound near the pole, wherever Subcarboniferous limestones occur, they carry essentially the same species of Crinoids and Corals and Shells, showing that the ocean from equator to pole was everywhere essentially of the same temperature. From Alabama to Pennsylvania, from Pennsylvania to the seventy-eighth parallel in the Arctic Zone, in

Asia and Europe and both Americas, wherever coal-beds occur, we find them composed of the same types of vegetation, showing that in torrid and temperate and frigid zones the air held essentially the same temperature. It was warm and moist and heavy with the gas of carbon. An atmosphere loaded with carbonic acid, as experiments have shown, would be no bar to the sun's rays as they fall *on* the earth, but would be as an overarching mirror to throw them back as they radiate *from* the earth. Warmth increases evaporation, and the carbon-laden air must have been laden with mist. A dull, heavy, foggy air enveloped the world and veiled from the sun the dense jungles of Tree-fern and Seal-tree. Their degenerate kindred of to-day love shade as well as moisture.

Ocean, and air into which it poured its steaming vapors, and half-risen land over which it returned now and then to assert its dominion — that was the world in times Carboniferous. The pulses of nature were slow. Sluggishness was everywhere. Ocean and air were tranquil, for differences of climate in which storms are born, were hardly known. Streams were sluggish, for the land level was hardly above the sea. The mailed Amphibians were sluggish, for they were cold of blood and they breathed an air which no animal quick of pulse could breathe and live.

The world was not good but it was in process of becoming good. The Fern forests were purifying the air by taking from it its carbon, and preparing coal for far-off ages by storing the carbon in the rocks.

The world was young, and yet, if man could have breathed its air and waded through its jungles and strolled along its sandy flats where land blended with sea, he would have said "it is growing old." For he would have seen indications that the great Scale-trees and Seal-trees were fading out and were soon to become extinct. Looking into the sea he would have found that the Crinoids, having reached their culmination were beginning to wane. He would have seen that the Spiral-gilled Mollusks* were dying out, and that the cup-leaf corals, (Cyathophyllum) spread once over every reef, were almost gone. Turning then to the rocks he would have found registered there the history of dynasties long extinct. Like us he would have gazed with wonder on the Trilobite and theorized as to the affinities of an animal, in pattern so remote from anything in his "modern world." He might have looked on the same delicate plume-like fossils *we* have seen in the lower rocks of the Lower Silurian, and deciphered their affinities with living Hydroids and indulged in speculations as to the time, in a scale of years, since the class of Graptolites became extinct.

Turning now from the sea and the rocks to the jungles, and pondering deeply the meaning of what he saw, he would have found in this monotonous repetition of stems, fluted and scarred and set with spiky leaves or crowned with spreading fronds, the fore-growth of groves beautified with flowers and enriched with fruit, and in the shed fronds and fallen stems at his feet he would have found the promise of warmth and light for man in ages as far in the future as that of Eozoon in the past. Looking from the jungle to the low slopes on its borders. one feature

* Worms, in reality.

of the coming time would have stood before his eyes, the Pine of the North and the Palm of the South.

Progress there had been, and was to be. Can we reconcile our thoughts to its slow pace? Why did the world, having got a plant, wait so long for a flower? Progress was conditioned, in part, on the air. Perhaps it was conditioned in larger part on the light.

Mr. Sorby of Sheffield, has made experiments on the effect of different vibrations of light on different orders of plants. The experiments seem to show that short and quick-beating pulses stimulate the growth of flowerless plants like Ferns and Mosses, and that the longer and slower pulses have more potency to stimulate and build up the flowering plants. Now of the thrills of light from a sun in its youth, more are short and quick, while an older sun pours forth into the ether more of the longer and slower vibrations. White stars, like Sirius, bathe their attendant earths in floods of quick-pulsing light and may robe them in a livery of Lichens and Moss and Ferns. According to the teaching of Sorby's experiments not a new-created earth circles around a new-created sun which the hand of the Maker has gemmed with flowers. If these teachings shall be borne out by other and fuller experiments they will explain the earth's slow progress from the Lichen to the flowering and fruiting shrub. The earth must wait on the sun — must wait till his cooler disc pours into space longer pulsations of ether, before flowers could gem the ground and fragrance fill the air. If a naturalist may find "the progress of the ages marked in the tails of Fishes," and in the sequence of Reptile to Reptile-fish, and

Stem-plant to Thallus-plant, a chemist may find the conditions of that progress far away in the sun.

In the sunbeam he *does* find the agent of that Power his analysis can never reach, the commissioned agent for the creation and conservation of all things. Through the æons of globe-history its nimble fingers have woven the tissues of life and garnered the elements freed by death. The forests it reared in fluted stem and spreading frond over the primeval fen, are stored in the rocks as beds of coal. The gardens of Algæ it spread as a fringe along primeval shores, and the gardens of Polyps it made to bloom under primeval seas are garnered in the globe's strata as rock-oil. As grains of wheat which went down into the tomb in the hand of Belzoni's mummy, three thousand years ago, come up now to swell and spring into new life, so the Fern, the Alga, and the Polyp, that went down into their rock-tombs millions of years ago come up now at the bidding of man to live again in the light and heat which wrought them.

"To be or not to be"

is not the question. Man or thing, whatever is, must always be. *Change* there is, but no destruction. And whatever is, whether plant, animal or man, it is sun-built. In the orient an Emperor calls himself the "Son of the Sun." We all sit on the same throne as that claimed by the oriental. We are all Sons of the Sun. And the bird and the beast, they are Sons of the Sun. And the weed and the worm, they too are Sons of the Sun. And the coal and the oil held in store by the rocks, even they are members of this vast celestial brotherhood. All, all are Sons of the Sun.

Wind and rain and steam and storm and light and life are modes of one power, lodged in the beams of sun and star, forever changing but never lost, a flux of energy eternally the same, rolling from the burning suns through the universe, rolling in rhythm through the geologic æons.

# CHAPTER IV.

The Cephalopod — Evolution of the Nautilus — Evolution of the Ammonite — Ontogeny and Phylogeny — The Individual and the Race — A Species in its Rise and Culmination — A Species in its Decline and Death — Evolution of the Saurian — Ichthyosaurus — Plesiosaurus — Ceteosaurus — Megalosaurus — Hints of the forthcoming Mammal — Evolution of the Bird — The Series which led from Reptile to Bird — The Pythonomorphs — Elasmosaurus — Protostega — Portheus — Pterodactyle — Pteranodon — Europe in the Cretaceous Epoch — America in the Cretaceous Epoch — Hesperornis — Bird of the Evening of Mesozoic Time.

IF WE build a fire on a sea-beach in any dark autumnal night, Mollusks in which the molluscan structure is masked under the quaintest of forms will be attracted by the glare and will throw themselves against the lighted sand. Sailors call them "squids." We place on Fig. 35 a member of this family called an Octopus. Hugo's "Devil-fish" was an Octopus. A short, cylindrical body with glaring eyes and eight long, flexible limbs surrounding the mouth — this is Octopus. The idea of a Coddington lens might have been taken from its eye, and that of an air-pump from the sucking cups which are set in double rows along each of the snaky limbs. A more common member of this family is the Squid. It has ten arms, eight of which surround the mouth in a circle and are of equal length. The other two are much larger and are attached, one on each side, within the circle of eight.

The inner face of each of the short arms is covered from base to tip with a double row of suckers. When the Squid clasps his prey, these suckers adhere with great tenacity. The Squid has two gills, in shape something like leaves. It has a rudimental, horny

Fig. 35.

shell, which lies inside of the body and is shaped something like a pen. In the Cuttle-fish, another member of the family, this rudimental shell, which lies in the same position in the mantle, is calcareous and spongy, and is known in the market under the name of "Cuttle-bone." In the Nautilus, a rare

Fig. 35. Octopus fulvus.

member of the family, the shell is no longer a rudiment and no longer internal, but large, pearly, coiled and chambered, and external to the body. The gills are not *two*, as in Octopus and Squid and all other naked members of the family, but *four*.

The position of the limbs (they are on the head) gives us a name for this highest order of Mollusks. It is *Cephalopod.* And as the number of gills, whether two or four, is found to be in correlation with other and important structures, we will divide the order into two sub-orders, *Dibranchiates* and *Tetrabranchiates.* All naked Cephalopods are dibranchiate; all shell-bearing Cephalopods are tetrabranchiate.

Emerson has said that a crab by the sea-side is commonplace, but a crab lifted up into the heavens and made a sign of the zodiac is sublime. Squid and Cuttle-fish are commonplace. We must lift them up and make them signs to map the zodiac through which creation has moved.

The first recorded Cephalopods were in the Old Silurian sea. They are known by the shells they have left in the Silurian rocks. Shells are the real "medals of creation." A fossil bone registers the animal at the stage when death overtook it. A fossil shell registers each stage of growth up to the period of death. Shell-bearing animals are known by their shells in a fuller sense than Mammals or Reptiles or Birds by their bones.

The shells of the Cephalopod appear first as long, conical, and chambered. They are known as Orthocerata, a name whose English is "Straight-horn." Some of these shells were very long. In the Trenton limestone of Lowville, New York, a "Straight-horn" has been found more than eighteen feet long. The inmates of these shells lived in the outer chamber, and crawled like immense spiders on the bottom of the sea. Their tentaculæ were numerous and short, and as mere repetition of parts is a badge of inferiority, we must infer that their rank on the Cephalopod scale was low. We represent at No. 1, Fig. 36, one of these Straight-horns as it appeared crawling over the sea-bottom. With its unwieldy shell rising above it like a tapering column, eighteen feet high, it must have been sluggish and awkward. Its form was bad and could not endure. Already we find members of its family with the shell slightly coiled. The point of departure from the "Straight-horn" is shown in the Curved-horn (No. 2).* The next step was taken in the Small-mouth horn (No. 3),† and the next in the Ring horn (No. 4).‡ The species represented in Fig. 36 has the first three whorls coiled in contact. From straight shells we have come to discoidal ones. From discoidal shells like Lituites (No. 5) we come by another series to discoidal non-involute shells (No. 6), and from these to discoidal involute shells like the Nautilus (No. 7).

THE STRAIGHT HORNS,

and the like, are aberrant Nautili. The involute, discoidal Nautilus (6) was the most convenient form, and toward this creation was moving, from the Straight (No. 1), through this series of curves, to No. 7. The discoidal Nautali began in the Upper Silurian, in two species. The type unfolded until, in the Carboniferous

* Cyrtoceras. † Phragmoceras.

‡ Gyroceras. See numbers 2, 3, and 4 of Fig. 37.

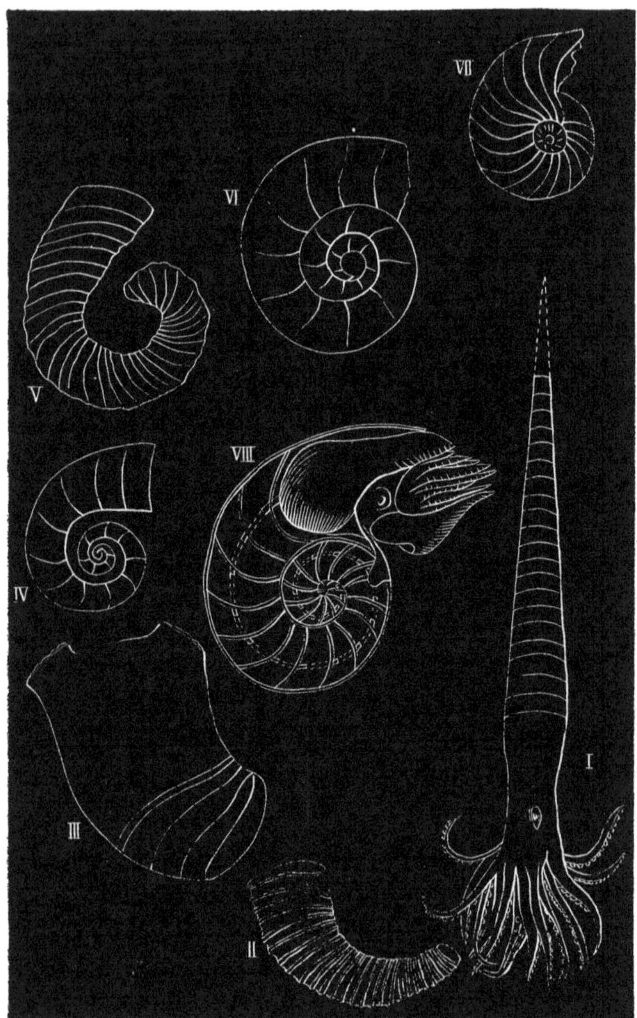


Fig. 36. **Evolution of the Nautilus.**—1. The Othoceras. 2. The Cyrtoceras. 3. The Phragmoceras. 4. The Gyroceras. 5. The Lituites. 6. Nautilus Konickii. 7 and 8. Nautili.

Fig. 37. Evolution of the Ammonite.—1. The Goniatite. 2. The Ceratite. 3. Ammonites margaritatus. 4. Jason Ammonite. 5. Scaphite. 6. Crioceras. 7. Ancyloceras. 8. Hamite. 9. Toxoceras. 10. Baculite.

period, it culminated in thirty-eight species. From the Carboniferous period it has been declining until, in our own period, it numbers only four species. The aberrant Nautili numbered a hundred and fourteen species in the Silurian. They declined to sixty-four in the Carboniferous, to four in the next period, and in the Jurassic they had died out altogether. The Nautili of the Carboniferous are not involute. The modern involute type of Nautilus was reached in the Jurassic period.

A tube called a siphuncle passes through the chambers of the shell, and in the Nautilus family it passes through the center, or near the center of the chambers. In another family called *Ammonites*, the siphuncle extends along the *inner* side, close to the shell-wall, or along the outer side.

The Ammonites began in the Devonian period, in the closely-coiled-discoidal *Goniatite** (No 1 of the second plate, Fig. 37). Its partition-walls, or septa, were flexed on the margin. As time passes the Goniatites become more and more complex in the flexures of their septa, and pass through the Ceratites (2) into the true Ammonites (3). The type unfolded into two hundred and twenty-two species, and culminated in the Jurassic. It began then to decline, and finally it died out in the Cretaceous.

Mere description and cataloguing of species are of little worth unless fertilized by an idea. The investigations of D'Orbigny and Hyatt and Würtenberger have set the history of the shell-bearing Cephalopods in the light of a profound philosophy.

* Gonia, an angle.

An individual lives not its own life merely, but it lives over again the life of the race to which it belongs. The life of the primeval man, with no history behind him, was simple. The life of a man who had the history of a few generations only to remember and ponder was not very complex. The life of a man now, whose mind would traverse the history of his race through all the generations behind him, is exceedingly rich and complex. Analogous to this increasing complexity in the intellectual life of man with the increasing age of his race, are the phenomena displayed in the *physical* life of an animal. The earliest shell-bearing Cephalopod registered in the rocks, the "Straight-horn," developed simply. The old was only a magnified copy of the young. The young shell had the same outlines and the same form of septa as the old. In its development it did not take on and throw off historic forms and structures. It is as if there was no history for it to repeat. Leaving this primitive pattern of chambered shell, we pass far down along the geologic ages till we find the Ammonite. We find it foreshadowed by the Goniatite, whose septa were bent along their margins into lobes. Passing the Goniatite, we find a nearer approach in the Ceratite, whose septa were bent into *pectinated* lobes. Coming at last to the Ammonite, we find the margins of its septa crimpled into a labyrinth of curves. Back of the Ammonite there is history. See now how complex its own life-history. A very young Ammonite is a reminiscence of a Goniatite. Its septa have four rounded lobes. Later on in its growth it seems to live over again the Ceratite. The lobes of its septa increase in number and become *pectinated*. When it reaches maturity the lobes of the septa have increased to eighteen and become *foliated.* It is now an Ammonite.

The analogy does not end here. An old man is for the second time a child. He is a "wrinkled infant." Look now at the old age of the Ammonite type. The last of the race was the Baculite, No. 10. The shell was uncoiled. It had the same form in age as in youth, and the septa had the same number and form of lobes. The septa of the Baculite, in youth and maturity alike, corresponded to the immature septa of the Ammonite. At the beginning of this family was the Orthoceras whose characters, through life, were infantile. At the end of the Ammonite branch of the family was the Baculite whose characters again were infantile. Baculite was an infant but *an old age infant.* In it the type has reverted to a primitive form and structure and was dying.

The analogies between the life of the individual and the collective life of the race are not yet exhausted.

The geologic record shows four periods of great expansion of shell-bearing Cephalophods. They are known as the Silurian, Carboniferous, Jurassic and Cretaceous. Hyatt has shown that these epochs may be distinguished as,

*First.* An epoch of rounded, unornamented shells with simple septa.

*Second.* An era of ornamentation.

*Third.* Zenith of development.

*Fourth.* Retrogression.

When the type had reached the zenith of development its ornamentation was most conspicuous.

Keeled, channeled, striated, ribbed, tuberculated and spined, forms abounded. The Keeled Nautili disappear, then the striated and ribbed, and now the few remnants of the type which remain are smooth. There are four periods in the life of an individual Nautilus. In the first the embryo is smooth and the abdomen or ventral portion of the shell is round. In the second the shell covers itself with tubercles around the umbilicus, or initial point of growth. In the third it covers itself with ribs, striations, and tubercles on the abdomen. In the fourth it degenerates and retrogrades. Its ribs and striations disappear, then its tubercles, and in old age it is without ornamentation and perfectly smooth.

The life-history of an Ammonite shows the same analogies. During the rising period of the order the ornamentation is increasing and the septa are growing more complicated. After the vital energies have passed the maximum, the declining period of the order commences, the whorls become less involute and more tubular, the septa become less foliated, and all the parts revert toward the simplicity manifested in the earliest members of the family. In the rising and zenith periods of the order the size of the whorls added in old age does not diminish. In the period of decline, while the order is passing through its old age, the radii of the whorls diminish in the old age of each member of the order. As the mutations of the septa are in relations with the form of the whorl, senility invades the septa and reduces them to simpler forms. Contraction of the whorl indicates spent vitality. It is an age character. See now how age creeps over a race as over an individual.

In the earliest Ammonite we find no age characters, such as contraction of the whorl and reduction of the foliations of the septa, even in old individuals. We pass upward through the strata and find old Ammonites of a later geologic period showing characters of age on a portion of the last whorl. We pass upward into strata still younger and see these characters creeping inward and invading the next whorl. In Ammonites still later in time we see these characters creeping in still further and invading the inner whorls. In the last Ammonites we find them pervading all the convolutions even to the inmost whorl. The shell "is written all over with the characters of age." It is as if you were to find man in one century ageing, let us say, at sixty; in the next at fifty; in the next at forty; and so on, earlier and earlier, until you find him *old even in infancy*. You would say that our race was nearing its latter end. The Ammonites of the Cretaceous period show that their race was nearing the end. As the last indication of senility the closed spiral begins to relax. At first this character like other age characters affects only the outer whorl. In the Jason Ammonite, No. 4, the last chamber is prolonged at a tangent to the convolutions. In the Skiff Ammonites, No. 5, the last chamber is prolonged still more and it bends on itself. In the Ram's-horn Ammonite, No. 6, all the whorls are separated from each other, but still the discordal form is retained. The Handle Ammonite, No. 7, begins as a Ram's-horn and terminates as a Skiff Ammonite. In the Hook Ammonite, No. 8, the process of uncoiling is carried still further. In the Bow Ammonite, No. 9, the shell is a segment of one loose, unsymmetrical coil. In the Ptychoceras the shell has become a straight shaft with a crook at one end. Finally, in the Cane Ammonite, or Baculite, No. 10, the shell is completely uncoiled. This is the last of the Ammonites, and it is perfectly straight like the first of the Nautili. It came in the extreme old age of the type and its age characters are so emphasized that it represents the second childhood of its race.

Wonderful, in the light of science, is the life of the humblest thing on this planet of ours — wonderful the life of the lowliest of its tenants in the dimness of the past; and wonderful too, if we ponder it, is death, death which has made the planet's crust a vast sepulchre of its cast off races. Wonderful the individual life—life rising and ornamenting itself in the individual as it rises and unfolds itself in the order, declining and dropping its adornments in the individual as it declines and drops its adornments in the order, and pulsing in one Ammonite through the span of a dozen years in the same rhythm as in the collective race of Ammonites through the span of a dozen million years. Wonderful too the law that makes the ageing and dying of the individual epitomize the decline and death of the order. For we have learned from the history of the Ammonite that law presides at the death-end of a species as well as the birth-end. The law must be of wider application, although the details have not been worked out. It should apply even to the races of men. When age-characters appear early and habitually either in the skeleton or in the mind of the members of a race, that race must be on the way to extinction.

The history of the tetrabranchiate Cephalopod has

been most instructive and suggestive. We have seen the earliest members of the order repeating no ancestral history in their development, and creeping over the sea-bottom, carrying youth into age, and burdened with long, conical shells. The shells being badly shaped, we have seen them give way, little by little, as the ages advance, to coiled shells. We have seen the coils grow closer and closer with the passing ages, until the convolutions are *involute* and all the inner coils are hidden under the outer. We have seen these changes occurring so gradually as to lead to the belief that they were successive modifications and not new creations. In the Ammonite branch of the family we have seen ribs appearing first, then tubercles, then spines. In this order we have found the ornamentation coming on the individual, and in the same order we have seen it coming, through vast ages, on the race. In the growth of the individual, ribs pass into tubercles and tubercles into spines. In the unfolding of the class we find ribs pass gradually into tubercles and tubercles into spines. The mind is forced into the conviction that the spined Ammonite of a later age is an outgrowth of the ribbed Ammonite of an earlier age. The history is suggestive of the genesis as well as the death of species.

Another dynasty which appeared before the close of the palæozoic age, unfolded through the mesozoic age and declined with its decline, is that of the Reptile.

Reptiles have cold blood, like Fishes. Their temperature differs but little from that of the medium in which they live. Unlike Fishes they have true lungs and breathe the air. But a class of Reptiles called

Amphibians, while young, breathe by means of gills, and some of these Amphibians retain through life the gill-openings but not the gills, while others retain even the gills.

As Reptiles reach down toward the Fish in the gilled Salamander, so in other forms they reach up toward the Bird. Reptiles resemble Birds in having the two outer ear-bones external to the skull. These bones form in each class a part of the lower jaw and the bony rod which connects the jaw to the skull.

At the dawn of reptilian life the seas were swarming with reptile-like Fishes. They were gigantic in size and predacious in character. Their huge teeth were conical and striated like those of Saurians. The sutures of the skull were close as in Saurians.

As the Fish of carboniferous seas had affinities with the Reptile, so the Reptile of carboniferous marshes had affinities with the Fish. Some of the Labyrinthodonts had sculptured bony plates on the head like Ganoid Fishes. Many of the early Reptiles had bi-concave vertebræ like Fishes. But the horizon line on which Fishes and Reptiles meet we do not find.

From the Carboniferous system which records the first known Reptile we pass up to the Permian in which we find the *Proterosaurus*.*

The Permian Reptiles had bi-concave vertebræ, and teeth inserted in sockets like Crocodiles.

From the Permian we pass up to the Triassic which records the Reptiles of the period in a very curious way. In America we find hardly a trace of Triassic Reptiles except in their footprints. In England we find only remains of their remains. The sand had

* Proteros, first, and Saurus, lizard.

accumulated around the bodies of Reptiles, and hardened. After this had occurred water percolated the porous rock and dissolved out the bones, leaving nothing but the cavities. Thus we have only fossils of fossils, or remains of remains; but casts taken from these cavities enable the skilled anatomist to make out the most important characters of the lost Reptile. The Crocodile appeared in Triassic times, and we learn by casts of the cavities which held its remains, that in certain anatomical features it approached the Lizards. The Saurian preceded the Crocodile, which stands higher on the reptilian scale and which is a modification of the Saurian type.

We have come now to the mediæval period of the earth's history. We are in the

AGE OF THE REIGN OF REPTILES.

In the Jurassic period, which followed the Triassic, the reptilian type underwent a remarkable expansion. The Reptile was everywhere and lord of all the elements. It was mounted on paddles for the sea, mounted on pillar-like limbs for the land, mounted on bat-like wings for the air. The Enaleosaurs* are of higher grade than the sea-saurians of the Triassic. The advance is marked chiefly in the skull which in Triassic times had but little bone and which now in Jurassic times has solid bony walls.

One of the greatest Enaleosaurs was the Ichthyosaur,† No. 4 of Fig. 38. The skull was remarkable for the great elongation of the snout. It is found even eight feet long. It was remarkable too for the enormous eye. In some of the largest skulls the eye-

* Enaleos, marine. † Ichthys, a fish.

socket is three feet in circumference. A circle of bony plates was developed in the sclerotic coat. Such plates are found now in the sclerotic of certain Birds and Turtles. The maxillæ are reduced to small, slender, rod-like bones, as in Birds. The vertebræ are deeply bi-concave as in Fishes. The ribs begin just back of the head and continue almost to the end of the body. In the region of the tail they are short and straight, but in the precaudal region they are long and curved. They are attached to the vetebræ by two forks into which they divide and which unite with corresponding tubercles on the vertebral segments.

This is a structure found no where else in nature. The scalpula is long and narrow, as in Lizards, and the coracoid, or "crow-beak" process, is broad. At their junction they give rise to a cavity which receives the head of a short, prismatic humerus. The further end of the humerus, called the "distal end," presents two facets which articulate with two short, flat, polygonal bones, representing the radius and ulna. The carpus is represented by two rows of small polygonal bones, and the metacarpus by four quadrangular ossicles, following which are three or four or five series of polygonal bones representing the digits. An apparent multiplication of the number of digits, according to Prof. Huxley, arises from the occasional bifurcation of some of the digits and by the superaddition of marginal bones which may represent the remains of digits in a polydactyle fin of some ancestral fish. The paddle formed of these elements is unlike the paddle of a Turtle or Whale.

The bones of the hind-limb have no connection with the vertebral column. This limb has essentially the same structure as the fore-limb, but is smaller, and in some species much smaller, in size.

In anatomical structure the Ichthyosaurus was compounded of Fish, Saurian, and Bird, with peculiarities in vertebræ and ribs and limbs, which give it no place in any order, living or extinct. In shape it was something like a Whale. Like a Whale it had an enormous head which passed at once into a body forty or fifty feet long, propelled by large leathery paddles, and tapering into a tail which may have aided propulsion by an expansion of its integument into a fin. Unlike the Whale the enormous head had an enormous eye.

Another paddled monster was the Plesiosaurus (No. 5, Fig. 37). A little, snake-like head was mounted on a long, swan-like neck which sprang from a short body mounted on four large paddles, and terminated by a short tail. Such a quaint form had Plesiosaurus, a form adapted for swimming and fishing in bays and estuaries. There were many species, and the type extended through the mesozoic age from the Trias to the Chalk. The earlier species differed from the later chiefly in the shoulder-girdle, and this difference was very significant in kind. A curved bar of bone, consisting of three united pieces, connected the scapulæ. From the constituents of this bar came the interclavicles and clavicles of later reptilian species.

It would seem that many of the Saurians and Fishes found fossil in the Lias division of the Jurassic must have perished through some catastrophe. Dr. Buckland writes that scarcely a bone or scale has been removed from the place the animal occupied during life, which could not have happened if the bodies of these Saurians had been exposed to the attacks of Fishes on the bottom of the sea. Skeletons of Ichthyosaurs are found entire, and even the contents of their stomach are found between their ribs. From these undigested contents we can determine the character of their food, what particular fish they preyed upon, and from their coprolites we can tell the peculiar structure of their intestines. The spiral markings on the coprolite of an Ichthyosaur tell us that the intestine was spiral like that of a Shark, and give us one more point of affinity which one of the oldest of Reptiles had with one of the oldest of Fishes.

Lyell has called attention to the fact that multitudes of dead Fishes are seen floating on the sea after the discharge of noxious vapors during an earthquake. To such a catastrophe we may ascribe the sudden extinction of multitudes of ancient Fishes and Saurians.

The next member of the Jurassic system above the Lias is the Oolite. It is so named because it is composed of small, egg-like grains, resembling the roe of a Fish. Each grain is a concretion of lime around a nucleus, usually a small fragment of sand.

In the upper belt of the Oolite we find indications of approaching land. The indications are ripple-marks and bands of bituminous shale.

At the close of the Oolite period what is now the "District of the Weald," in England, was the bed of an estuary. On the floor of this estuary were spread out beds of clay and sand and limestone. The small deposits of limestone in this formation are of fresh-water origin. In these estuary deposits, called Wealden, the remains of land animals accumulated.

Bodies from the adjacent land, borne down by floods, sank into the sediment and were petrified.

The land whose eroding rivers flowed into the sea of the Oolite and the estuary of the Wealden has become classic as the land of the Iguanodon and Megalosaur and Ceteosaur. A thigh-bone of Ceteosaurus is five feet and four inches long, and at its upper extremity five feet in circumference! And yet this enormous thigh-bone shows by its mode of growth —it ossified from one point—that it belonged to a cold-blooded, egg-laying Reptile. The caudal vertebræ were bi-concave and the dorsal were convexo-concave. It is as if the monster were Reptile in its back and Fish in its tail. Its length was about fifty feet, and its height about twelve. Ceteosaurus was what its name implies, the *Whale*-saurian, the largest land animal, perhaps, that ever appeared on the globe. It seems to have been herbivorous.

Another gigantic vegetarian was the Ignanodon. Its length was about thirty feet, and its height about six. The femur was thirty-three inches long and the humerus nineteen, making the fore-limb a little more than half the length of the hind-limb. The teeth were broad grinders serrated on the edges. The front of the upper jaw was toothless and curved down like the beak of a Parrot. The faces of the vertebræ were nearly flat. These structures are taking us away from the Reptile toward the Bird and the Mammal.

The Megalosaurus will take us further on the mammalian line and as far on the line toward the Bird.

Wading through the shallow waters, or prowling along the marshy shores, was the most gigantic of carnivorous Saurians. The teeth of Megalosaurus were large, curved, and crenulated. A tiger is not toothed so fiercely as Megalosaurus was toothed. The limb-bones were hollow, as in Birds. The sacrum was composed of five vertebræ, as in Mammals. The pelvis, hind-limb, and foot were transitional between the

Fig. 38.

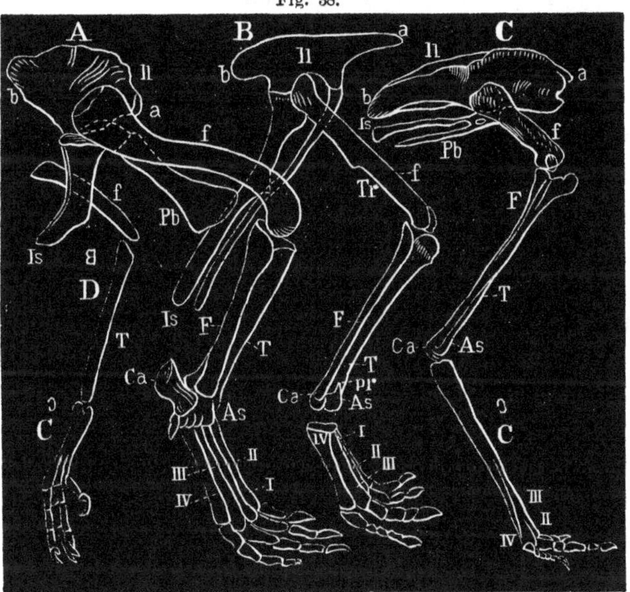


Reptile and Bird. Fig. 38 will show this better than words. We give three cuts after Huxley—A, of a Crocodile; B, of a Dinosaur, and C, of a Bird. A glance at this cut will show that in the Reptile the ilium (*il*) is not prolonged in front of the cup which receives the thigh-bone, that in the Bird it is greatly

Fig. 38. Evolution of the Bird.—A. A Crocodile. B. A Dinosaurus. C. A Bird. D. The Bird-reptile of Solenhofen. Il. Ilium. Is. Ischium. Pb. Pubis. f. Femur. F. Fibula. T. Tibia. Ca. Calcaneum. As. Astragalus. 1, 2, 3, 4. Metacarpal bones.

prolonged, and that in the Dinosaur it is prolonged, though not so much as in Birds. The ischium in Reptiles (*is*) is an elongated bone which extends downward, inward, and backward. In Birds it is elongated and inclined backward. In the Dinosaurs it is greatly elongated, and in Iguanodon it is even longer and more slender than in Birds. In Reptiles the pubis (*pb*) is inclined forward and downward. In Birds this bone is directed downward and backward parallel to the ischium. In some of the Dinosaurs it is in line with the ischium, and slender and elongated. A comparison of the limb-bones will show that those of a Dinosaur were nearly intermediate between Reptile and Bird. More significant than the limbs are the feet.

In the Crocodile the heel-bone (*ca*) is large and long. In the Dinosaur it is reduced to a little cap on the fibula. In the Bird it is gone. It has united with the fibula, which, in turn, has united with the tibia. In the Reptile, the astragalus, or turning joint (*as*), is a large bone with pulley-like surfaces for articulation with the bones above and below. In the Dinosaur this bone (*as*) is small and slender, and it throws up a little process (*pr*) over the tibia. In the Bird it has gone. In the Ostrich, which is the most reptilian of Birds, the astragalus remains for some time distinct, but in other Birds it unites with the tibia while in embryo. The skeleton of the Bird represents the extreme of anchylosis. The bone (*T*) results from the soldering together of tibia, fibula, calcaneum, and astragalus. In the Reptile all these elements are distinct and strongly pronounced. In the Dinosaurs the fibula is weak and slender, and the calcaneum and astragalus are little better than vestiges.

It is a very significant fact that while in Iguanodon and Megalosaurus and other Dinosaurians the astragalus remained through life a distinct though atrophied bone, in other members of the family nature took another step toward the Bird and soldered the astragalus to the tibia.

The metacarpal bones (1, 2, 3, 4,) are soldered together in the Bird and form the cannon-bone (*c*). The little toe is suppressed both in Reptiles and Birds, and the normal number of digits is four. In many Birds the hallux, or big toe, is suppressed, and of the three digits left the middle one is always the longest. Now in the Dinosaurians the metacarpals are elongated and fitted together so snugly that they can hardly move, the one on the other. Here is another step on the road that leads from Reptile to Bird. The metacarpals are not yet anchylosed into a single bone, but united rather into a *bundle.* The next step was taken within the limits of the class. In the Solenhofen slates the ruins of a wonderful little Saurian have been found. We give at D a representation of the limb and foot of this Bird-saurian. Tibia, fibula, calcaneum, and astragalus are completely anchylosed in the bone (*F*). The metacarpals are anchylosed in the bone (*C*). The digits were four. The femur (*f*) was shorter than the tibia (*t*), another bird-character, as a glance at the series will show. In limb and foot the transition was complete and the Reptile had become a Bird! In the form and size of the head the transition was complete, but the Reptile lingered in the teeth and in the tail.

These details may have been tedious, but they have

shown the path along which creation has moved. Birds came from Reptiles, and the Dinosaurians represent nature in the transition.

For Megalosaurus we draw a hind-limb, large, long, and terminated in a four-toed, bird-like foot. As the fore-limb was short and weak it could not have been used as an organ of locomotion, and our great Saurian

Fig. 39.

must have walked on two legs. If nature would poise a horizontal body on two legs she must place the viscera backward and lighten the anterior regions ot the body. For Megalosaurus, therefore, we should draw a comparatively small head and neck and fore-limbs; as the animal had strong carnivorous teeth, it must have had hooked claws, as such teeth always imply such claws.

Fig. 39. Megalosaurus.

Following the Jurassic period was the Cretaceous. Chalk is a deep-sea rock, and is simply a compact, comminuted mass of minute shells. The Cretaceous system has an immense development in America, but, except a few patches in Kansas, it contains no chalk. In this system we read the last chapter of Mesozoic history. Our reading shall be from the Cretaceous of North America.

Between the Missouri River and the Rocky Mountain lies a vast plain, the bed of an ancient sea, overlaid here and there by the bed of a lake not so ancient. Here and there over this plain the rain-channels are deep gorges which wind about and open into each other and form labyrinths; and where parts of the intervening masses have been eroded and parts left standing in the form of pillars or turrets or spires or rounded domes or jagged walls, they have formed the *mauvaises terres*, or bad lands. The ancient sea was a Cretaceous sea, and in its limestones we are to find the wrecks of such animals as peopled it, and in the strata formed on the bed of its gulfs and estuaries we are to find wrecks of such animals as prowled over its adjacent lands.

Here, in a limestone bluff, is the stranded wreck of the greatest of Sea-turtles. The ruins proclaim a Turtle of prodigious bulk. The expanded flippers had a spread of more than fifteen feet! But the interest turns not on the bulk but on the structure. Prof. Cope who discovered this giant of the Cretaceous sea, has pointed out a significant feature in its make-up. It is well known that the shell of a Turtle is formed out of elements common to every vertebrate skeleton. The roof of the shell is formed by a union of the expanded ribs with bone deposited in the skin. The lower shell is formed of united ribs and breast-bone. In young Turtles the ribs are separated as in other vertebrates. As the Turtle grows the ribs begin to expand at their upper end. As time passes the expansion extends downward through the entire length. The bone deposited in the skin begins as ossicles laid down in independent centers. These separate pieces of bone expand and early in the life of the Turtle, unite with each other. Now the ribs of this great Turtle of the Cretaceous sea corresponded in form to the ribs of a modern Turtle when just hatched, and the bony plates deposited in the skin were free, as they are in the embryo Turtle of modern times. *Protostega* — for that is the name of the Cretaceous wonder — was a colossal *immaturity*. Other animals of the ancient world appeared in similar incompleteness.

And here is the wreck of another monster of the Cretaceous sea. By projecting our minds and putting them in company with Cope and Whitten, we live in ourselves the golden moments attendant on discovery. Passing a bluff of limestone, we see, projecting from its face, a portion of jaw and teeth. They arrest our attention, and attacking the bluff with pick and spade, we soon uncover them. They call up in the mind the image of an immense Python as they lie there, bristling, each jaw, with four rows of murderous teeth which paved the floor and roof of the mouth. More digging and we lay bare the vertebræ of the neck, then the pectoral arch, then the dorsal vertebræ and ribs, the pelvis and part of a hind-limb, and finally the massive tail which stretches away into the bluff. But the parts are displaced. Unlike the Ichthyosaurs, which lay undisturbed on the bottom of the Jurassic sea and were buried at once under sediments, when the body of this Saurian had sunk through the Cretaceous sea it was snatched and pulled and dragged hither and thither by predaceous Sharks. The monster was nearly a hundred feet long. What was it? Its head was large and flat, its jaws long, its eyes directed upward. A very anomalous feature was an articulating joint in each ramus of the lower jaw. The joint took the place of the usual suture. Its position will be seen at *a* in the restored lower jaw. Its action was

Fig. 40.

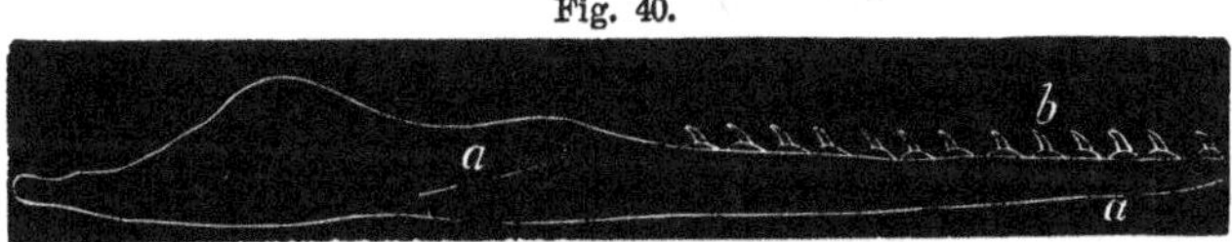


lateral. You will imitate the action of the lower jaw by holding your arm at full length and then flexing it outward at the elbow. As each ramus was free at the end, the creature had really *three* jaws, one upper and two lower; and you will imitate the lower jaws by extending both arms and then moving the elbows outward and the hands inward toward the breast. By this movement the python-like monster after seizing his prey, worked it down his throat!

The limbs and limb-girdles were small. The scapula was the segment of a disk, differing in form from that of any other order of Reptiles. The limb, composed of the same elements as that of a higher vertebrate, appeared in a form resembling the flipper of a marine Turtle.

The reptile was related to the Lizards and the Ser-

Fig. 40. Lower jaw of a Pythonomorph.

pents, but sufficiently distinct from each to constitute another order, the *Pythonomorpha*. This, the largest of Pythonomorphs, is called *Lyodon dispelor*. It was a genuine Sea-serpent, and glided through the Cretaceous seas something like an immense eel. It had companions as *outré* as itself.

Near the boundary line between Kansas and Colorado, on the escarpment of a bluff of clay-slate containing gypsum, Dr. Turner found the remains of a marine Reptile whose distinctive feature was a long neck and *flat* tail. The neck was composed of sixty or more vertebræ. The length of the neck was twenty-two feet and the entire length of the animal forty-five feet. In one species the neck was three times as long as the body. It rose from a body of elephantine bulk and terminated in a little arrow-shaped head. The body tapered down into a flat tail and was mounted on two pairs of paddles, like those of a Plesiosaurus. The flat tail gives a name to the order, *Elasmosaurus*. The Elasmosaur differed from the Plesiosaur chiefly in the arrangement of the breast bones. It was carnivorous, as the teeth testify, and the remains of Fishes found in the place of its stomach. An Easmosaur is shown at No. 3 of Fig. 41.

Hardly less interest attaches to the Cretaceous Fishes than to the Cretaceous Reptiles. When Armstrong doubled the destructive force of his gun, shipbuilders doubled the resisting force of their iron-clads. And when nature had developed such destructive powers in her Sea-saurians she must either drop her Fishes or preserve them by making them prodigiously fecund or prodigiously strong. She did both. What a prodigy of strength she made her *Portheus!*

The head was longer than that of a grisly bear and the jaws were deeper and stronger. The muzzle was deeper than that of a bull-dog. The teeth were sharp, cylindrical, glistening fangs, planted in pits an inch deep and projecting three inches from the jaw. Two pairs of enormous fangs crossed each other at the end of the snout. When Portheus met Elasmosaur or Pythonomorph, the Cretaceous sea was not a picture of that pre-Adamite innocence and repose which moralists ignorant of science have delighted to paint.

The sky like the sea was tenanted by Saurians.

From the yellow chalk of Butte Creek, in Western Kansas, were taken the ruins of the largest of flying Reptiles. Pterodactyles have long been known and have long excited the interest of naturalists. In the head a Pterodactyle approached the bird-type. The brain-case was more rounded and consequently more bird-like than that of any other Reptile. The occipital condyle — that rounded boss of bone by which the head articulates with the neck — is on the base of the skull, as in the bird. But the reptilian type is kept in the presence of a postfrontal bone, and a wide departure from the Bird is shown in the presence of teeth. The neck was long, the tail short, and the fore-limbs were much larger and stronger than the hind ones. The ulnar digit, or that which answered to our little finger, was greatly prolonged, and a leathery, bat-like wing was attached to it.

A great many species of these winged Saurians have been found in the Jurassic and Cretaceous of Europe, some not larger than a sparrow and others with a wing-spread of nearly fifteen feet. Only a few species are found in the Cretaceous of America but these were

10

the largest of the race. From tip to tip of the outspread wings, the largest Pterodactyle that flew above the Cretaceous sea of Kansas measured twenty-five feet. A new order has lately been found, named by Marsh, from the absence of teeth, *Pteranodon.*

Fig. 41. Kansas during the Cretaceous Period.

The land was, as the sea and the air, the abode of Reptiles. The Cretaceous beds of Colorado have yielded remains of Dinosaurians, in form something like Megalosaurus. There were species great and small, carnivorous and herbivorous.

We have been reading the records of an age removed from ours by millions of years. Over a great part of what is now Europe there rolled a deep ocean. The Alps and the Jura were forming as mud-sediments along the subsiding troughs of its bed. Europe was a great archipelago with its largest land area in the north. The Norwegian and the Ural were its only mountains. A great part of Asia was under a deep sea and the Himalayas were, as the Alps, still sediments along its subsiding troughs. In America the Appalachians and the Laurentians of Canada were up, but the sites of the Andes and Rocky Mountains were still covered by a mesozoic sea. The coral reefs which grew on the bed of the mesozoic seas show that the world was warm to the sixtieth parallel north of the line and to the Straits of Magellan on the south. A sea whose eastern coast we can trace from Arkansas nearly to the head of Lake Superior flowed over the great plains and far westward to a shore which science has not found. Along the shores of that sea were prowling the bird-like Dinosaurs whose ruins lie in the Cretaceous rocks of Colorado. On its broad expanse were disporting the colossal Turtles and Pythonomorphs and Elasmosaurs whose wrecks are lying now in the limestones of Colorado and Kansas. Over that sea, lazily flapping their leathery wings, were the great Pterodactyles and Pteranodons whose ruins are entombed in the same rocks as those of Pythonomorph and Elasmosaur. It was a world in transition to a better world. The reptilian dynasty was nearing its end. *Cold* blood must yield the scepter to warm. A herald of the new was already in the air of this very Kansas.

For a long time we have seen the coming of the Bird. We have seen it coming in *fragments*, here a pelvic arch, and here a limb, and here a foot, and here a rounded brain-case and horny beak. In the chalk-making age it came almost complete. In the

Cretaceous rocks of Kansas are found the remains of *Hesperornis*, a name whose English is "Evening-bird" — Bird of the evening of mesozoic time.

> "— last in the train of night,
> If better it belong not to the *dawn*."

Hesperornis had teeth — an heir-loom from the Reptile. Another Bird whose ruins we take from the same rocks, the *Ichthyornis*, had bi-concave vertebræ, an heir-loom which the Reptile inherited from the Fish and the Bird from the Reptile.

Here is a picture of creation drawn from the beliefs of a past generation:

> "The sixth, and of creation last arose,
> With evening harps and matin; when God said,
> 'Let the earth bring forth soul living in her kind.'
> . . . . The earth obeyed, and straight
> Opening her fertile womb teemed at a birth
> Innumerous living creatures, perfect forms,
> Limb'd and full grown: out of the ground uprose,
> As from his lair, the wild beast, where he runs
> In forest wild, in thicket, brake, or den;
> The grassy clods now calved; now half appear'd
> The tawny lion pawing to get free
> His hinder parts, then springs as broke from bonds,
> And rampant shakes his brindled mane."

Such a strain on such a theme!

> "Now half appear'd
> The tawny lion pawing to get free
> His hinder parts."

Contrast with this mud-birth of the lion as sung by Milton, the genesis of mammal and bird as indicated by science. "Now half appeared, in the pelvic arch of a Dinosaur, the Mammal; and now, in the neck and head of a kindred Dinosaur, there half appeared a Bird, *its hinder parts already free!*"

## CHAPTER V.

The Ice-age—Bowlders, Gravel, Clay, Sand, Planed and Scratched Ledges — Ice from Water and Ice from Snow — Glaciers of the Alps — Glaciers of Greenland — Motion of Glaciers — Glaciers Made and Moved by the Sun — Europe in the Glacial Epoch — America in the Glacial Epoch — Warm Periods Intercalated between Glacial Periods — Cause of Glacial Periods — Times of their Occurrence.

COMMON things, when seen in the light of science, are not commonplace. One of the most common objects before the eyes of a dweller or tourist in the north, whether of Europe, Asia, or America, is "the erratic bowlder." Detached rocks, having no kinship with ledges below, lie here and there over the surface of North America, from the polar regions to the latitude of Philadelphia. Here and there over New England they lie in bleak profusion. At Carver's Harbor, Mount Desert, they appear in such bulk and numbers as to arrest the attention of the idlest stroller on this summer asylum of the overworked. Along the stage-road between Bangor and Ellsworth, or between Ellsworth and Bucksport, huge blocks, many hundred tuns in weight, are seen perched on naked ledges, and perched, here and there, block on block.

> "Crags, knolls, and mounds, confusedly hurled,
> The fragments of an earlier world."

Putting ourselves in the attitude of tourists, let us

question these loose rocks and the clay and sand and gravel and rock-scratches associated with them.

On the north shore of Lake Superior our attention is arrested by ledges of "rose granite." The rose-color is in the feldspar, and a little attention to the characters of this granite will enable us to identify it wherever we may chance to see it. *In situ*—which means at home—we shall *not* see it except here on the northern borders of the great lakes, but traveled bowlders of it we may see over regions south and southeast of the lake. We pass to the south shore and find rocks veined with native copper. We continue our journey southward and pick up now and then in the gravel-beds or clay-beds a fragment of this very copper, which must have journeyed, like ourselves, from the Upper Peninsula of Michigan.

We make our way into Vermont and find, in Stamford, a mountain composed of a peculiar type of granite called "foliated granite." The mountain rises in the form of a truncated pyramid. Its head has been clipped off and carried away. *North* of this mountain we find not a bowlder, not a pebble of foliated granite, but *southward* we would find fragments strewn along the slopes of Hoosac, and beyond the Hoosac, over the face of Massachusetts, almost to the sea. If we were to climb Hoosac Mountain and stand over the Tunnel, we would see, perched on the very top, a bowlder seventy-five feet around the base and twenty feet high. It has no kinship with the rock below. It does not belong, then, to the mountain, but is a waif carried there from some ledge of its own kind. It is foliated granite, and the parent ledge must be sought in Stamford Mountain. From Stamford to Hoosac the way lies over Deerfield Valley which is thirteen hundred feet deep, but that immense bowlder has been torn from the side of one mountain and laid on the summit of the other.

We leave Vermont and pass into Connecticut. Along the shore of the Sound we observe the type of rock changing as we go from New London to New York. Passing back in the direction of New London, along the north coast of Long Island, we shall find bowlders corresponding in character and order to the ledges we had observed in Connecticut along the opposite shore of the Sound. At Riverhead are bowlders of trap and brown sandstone, and almost directly opposite are the trap hills of New Haven and the brown stones of the Connecticut Valley. Along the beach, at Huntington, are bowlders of magnesian limestone and gneiss, and nearly opposite, in Connecticut, are the magnesian limestones of Danbury and the gneiss of Norwalk.

We do not care for further details. We have facts enough to give us a generalization. It is this: the bowlders strewn over the surface have been detached from ledges lying to the north. Further details would show that bowlders have moved, in general, from the northwest, although over some regions the movement has been from the northeast, and over others from due north. And these details would show that, in general, bowlders have not traveled more than forty miles, although some have been carried many hundreds of miles. The Stamford bowlder on the top of Hoosac; bowlders of the Canaan limestones in Connecticut, lying on Goshen, one thousand feet above their native beds; bowlders from the low land north of Mt. Katahdin, lying on the sides of the mountain, four thousand four hundred feet above the sea and three thousand above the ledges whence they were borne, show that the transporting force could carry its freight *upward* as well as southward. The absence of bowlders south of the thirty-ninth parallel, except those which are merely local, shows that whatever that transporting force may have been,it found a climatic barrier which it could not or did not overpass. We shall need these facts again, but for the present we lay them aside.

Within the area of erratic bowlders we find a hard, compact clay, called, sometimes "bowlder-clay," and sometimes "till" and known in some localities as "hard pan." This is the oldest of the superficial deposits, in which is written a most wonderful chapter of the earth's history. Other superficial deposits are beds of sand, and beds of gravel, and gravel and sand,

Fig. 43.

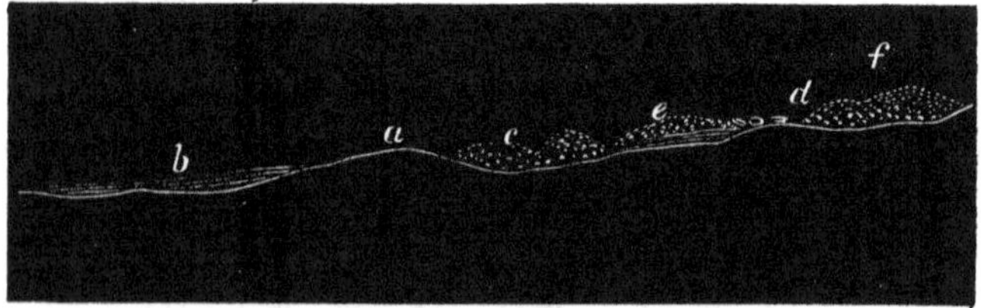


raised into ridges and rounded into mounds and hillocks. These surface deposits seem often a confused jumble, without order and without sequence. Sometimes a knoll of gravel will rest directly on a surface of rocks, sometimes on a bed of sand, and sometimes on the till. But in this seeming confusion there is order. The till never overlies the gravel. The section (Fig. 43) shows at *a* the fundamental rock,

Fig. 43. Ideal section of drift.

overlaid at *b* by till, at *c* by a gravel bed, at *d* by a knoll which may be sand or gravel, or both, at *e* by till which is overlaid by gravel, at *f* by a gravel bed over which rises a sandy or gravelly knoll. The section is ideal, but it expresses what might be found almost anywhere within the area of the drift. It shows the sequence of the surface deposits. It shows that the till is the oldest and the sandy or gravelly ridge the youngest. These facts will serve us when illuminated by others.

Here and there over the area of bowlder-clay and erratic bowlders we find abraded and grooved ledges. There is hardly a town in New England in which such ledges may not be seen. Limestone and slate and granite and trap have been rubbed and polished as a marble-cutter planes down and polishes his slabs of marble. Granite ridges and hillocks have been scarped and rounded and furrowed. Here is a picture of arctic scenery — the skeleton of the earth when the snow-robe is off — drawn by a recent traveler: "There was nothing to catch and detain the eye, which wandered on without a check over endless lines of round-backed ridges whose sides were rent into indescribably eccentric forms. It was like a stormy ocean suddenly petrified. Except a few tawny and pale-green lichens there was nothing to relieve the horror of the scene."

The grooves and scratches correspond in direction to the movement of the bowlders, as they correspond in area to the distribution of the bowlders. They were made, then, by the same force which transported these waifs. Now we find the rounded, planed, and scratched ledges, the traveled bowlders, the gravelly and sandy knolls, and the bowlder-clay or till, over the whole surface of North America, from the shores of the Arctic Ocean to the latitude of Philadelphia, and from the Atlantio coast to the longitude of Omaha. We find the grooved ledges and erratic bowlders from the sea-level to the height of five thousand five hundred feet above the sea. Some stupendous force has acted on the earth from the pole nearly half-way to the line, and from the sea-level nearly to the top of Mt. Washington. You stand on the bank of a river at low water and read in the stranded debris the record of a flood and the height to which it rose. We are translating the record of a mightier flood. We ascend Monadnock and stand on his bare and barren head. We place our hand on a naked ledge and find it as smooth as a lapidary could have rubbed it, and furrowed and scratched from north to south, as the burin of a graver would scratch a plate of steel. And here are bowlders which have no kinship with the gneiss below. They are waifs from the North; they are debris dropped by a flood. The flood overtopped Mt. Monadnock. For a measure of its height we will try Mt. Mansfield.

We ascend the highest of the Green Mountain Chain till we stand on its summit. Here too are planed and scratched ledges, and here too are travel-worn bowlders from the north. Mt. Mansfield was overflowed. As a measure against the deluge we will try Mt. Washigton.

Ascending the highest of the White Mountain Chain, by the carriage road, our eyes rest here and there on a bowlder which does not answer to the crystalline gneiss of the mountain. Near the "Half-way-house" we note the contortion of the strata and their planed and grooved surfaces. We pass over the summit to the Lake of the Clouds. We are now five thousand and five hundred feet above the ocean. Clouds are around us and below us. Far down lies a world dotted with sheets of purple water, and all around rises a world wrinkled into jagged mountains. We put our hand on a ledge and find it planed and grooved like those on Monadnock and Mansfield. We look around and see stranded bowlders, borne hither from parent ledges in the north. The abrading flood rose to the height of the Lake of the Clouds. We ascend still higher but find, not an abraded ledge, not a traveled bowlder. We have found at the Lake of the Clouds what seems the high flood line. The flood rose in the north, scarped and furrowed the round-backed ridges of Arctic granite, abraded the rose granite of Lake Superior, clipped the head from a mountain of foliated granite in Vermont, and from the shores of the Arctic sea down to the thirty-ninth parallel, and from the sea-level up to the Lake of the Clouds,

> "Heaped upon the cumbered land,
> Its wreck of gravel, rocks, and sand."

### WHAT WAS THE FLOOD?

If we take a handful of snow and round it and press it into a compact snow-ball we shall find that we have transformed it into granular ice. Ice which is pressed snow differs in many properties from ice which is congealed water. The hard pressed snow-ball will introduce us to the glacier.

If we ascend the Alps to the height of eight thousand feet above the sea-level we shall stand on an imaginary line called "the line of perpetual snow." Above this line, which in the Arctic zone is nearly at the sea-level, and in the torrid zone from thirteen to sixteen thousand feet above the sea, the water, evaporated and carried up by the sun, comes down always in snow, and falling on a mountain does not melt. Every Alp which rises above this imaginary line is snow-clad through all the year. Now as the snow does not melt and come down as water, if it did not slide off and fall down as an avalanche, every mountain which rises above that line would receive on its head an ever-growing mountain of snow. Through a cycle of five hundred years the sun lifts enough water from the earth and lays in the form of snow on the Alps to build on each Alp a snow-Alp a mile high. A time would come when all the lakes and even the great oceans would be lifted from their beds and piled up in vast Alps and Andes and Himalayas of snow. The wheels of nature would stop — blocked by snow. The solitude of a polar night so awful to the mind of Kane, would fill the world. But the snow must come down, and nature move on. We sat many hours of a summer day on the Wengern-alp of the Bernese Oberland. Till the sun had reached the zenith the stillness which reigned in these solitudes was oppressive to the mind. But the panorama was one of awful grandeur. To the left, beyond the valley of the Grindelwald, rose the flattened cone of the Faulhorn, to the right, the Mönch, the Eiger and the Schreckhorn, and in front, robed in her eternal shroud of snow, the Jungfrau. Around her base were immense fields of snow and snow-ice, while the Silberhorn and Schneehorn, her towering summits, shimmered in snow of spotless white.

The stillness was ended by a peal from the Jungfrau, echoing from peak to peak, till it broke on the silence like a succession of thunder peals. It was an avalanche. To the eye it seemed a little cascade of snow rushing down the Schneehorn and tumbling from precipice to precipice. But the ear is a better guide on these high solitudes than the eye. A snow cascade it was, but a cascade which might have torn a forest from its roots or whelmed a village in ruins. Such a cascade is an avalanche and the avalanche is nature's provision against the piling up of her oceans in Alps of snow. The fallen masses, becoming ice, move down to a valley and melt into water which flows into the ocean to be lifted up again as vapor and again laid as a mantle of snow on a mountain. So Nature moves on through her cycles and the pivot on which her vast machinery turns is the avalanche.

As a river drains from the land its surplus of water, so a glacier drains from the high mountains their reservoirs of snow. The fallen avalanches, and all the new-fallen masses of snow, press on the lower accumulations and transform them into granular ice, that is, into a glacier. From the Jungfrau to the Grimsel, from Grindelwald to Brieg, flanking the Jungfrau, the Mönch, the Eiger, the Schreckhorn, the Wetterhorn, the Agassishorn, the Finster-Aarhorn, the Lauter-Aarhorn, *all* the mighty mountains of the Oberland, and filling all the hollows between, is a glacier which covers an area of three hundred and sixty square miles. It is the largest glacier in Switzerland and is called the "Central Mass of the Finster-Aarhorn." It is a chain of connected glaciers one of which is classic ice to the man of science. A short walk from the Grimsel Hospice will bring the tourist to a small ledge of mica-slate jutting up from an ice-field and separating the Unter-Aar from the Ober-Aar Glacier. In 1827 the Swiss naturalist Hugi erected a hut under this sheltering ledge of rock and made it his abode for studies on the glacier. This hut in 1840 stood four thousand six hundred feet below the projecting ledge under which it was built, having traveled three hundred and fifty-three feet a year, or almost a foot a day. In 1841, Prof. Agassiz, accompanied by Forbes and Desor and Vogt, took up his abode in this vagrant hut and from time to time published bulletins dated from the "Hotel des Neuchâtelois." To the studies of these savans from their hut on the Aar Glacier we owe much of our knowledge of these moving ice-seas and their relations to the past history of the globe.

When the two arms of the glacier — Unter-Aar and Ober-Aar — meet and unite, the tourist will find a ridge of gravel, sand, and bowlders, which attains in some places, a height of eighty feet. Such ridges we call *moraines.* We see that a glacier can carry a freight of stones as well as huts. How does it receive such freight? Frost and ice are agents of destruction. Vapors penetrated the cracks and cleavage joints of rocks and freezing there they expand and act as a powder-blast. The water which flows into the joints of rocks, in freezing, will widen the joints and push the adjacent portions of rock further apart. This process goes on until blocks are detached from the ledge, and if the ledge is the side of a mountain with a glacier at its base, the detached blocks will fall into the glacier and ride away on its surface like Hugi's hut, or sink down into its crevices and by grinding against the ice and against each other, get themselves polished as they travel. The vapor that penetrates the little cracks and cavities of rocks, freezing and expanding, chip off fragments which roll down into the glacier and ride away on its face as sand or gravel. Such debris is piled up in ridges along the sides of the glacier, where they are called *lateral moraines.* When two glaciers meet and unite, as the two Aar-glaciers, their inner moraines unite and form a *medial moraine.*

The glacier takes in a freight from below as well as above. It abrades the ledges over which it moves. It crushes the rocks into gravel or grinds them into clay. It planes down the ledges, polishes them, and scratches them. Every bowlder, every pebble, imbedded in the under face of the glacier, is like a burin in the hand of a graver and scratches or furrows the rock-surface which the moving ice has abraded and polished.

A Swiss glacier has other lessons to teach us, but before we heed them we will do well to take a lesson from Greenland.

An old navigator has said that the sailor who sights Greenland finds nothing to cheer him in the sight. It is hardly so with the geologist. In that icy desolation he finds much which will help him to interpret a past condition of his own more favored zone. An "icy desolation" it is. Except a little strip along the western coast, Greenland is buried under a colossal mer de glace. The confluent glaciers of the Finster-Aar-horn cover an area of three hundred and sixty square miles. The area of Greenland is nearly eight hundred thousand square miles, and all this, save the narrow strip which faces an ice-choked sea on the west, is a lifeless solitude of snow and ice. The snow overtops the hills and levels up all the valleys, so that far as the eye can reach there is nothing but one vast, level, dreary expanse of white. Over all broods the silence of death. Life there is none, motion there *seems* to be none—none save of the wind which sweeps now and then, in the wrath of a polar storm, from the sea over the "ice sea," and rolls its cap of snow into great billows, and dashes it up into clouds of spray. But motion there is, activities we shall see there are, and on a scale of grandeur commensurate with the vast desolation itself. The mer de glace creeps outward and reaches the sea in tongues of ice which fill the indenting fiords. Eight hundred years ago there was a verdant strip along the eastern shore, and in the year one thousand the Danes planted a colony there. For five hundred years the colony maintained intercourse with its mother country; but the mer de glace, creeping seaward and detaching immense blocks, at last clogged the shore with ice and cut off access from the outer world. For three hundred years not a word has come from that ice-locked colony to us, and not a word from the great world without has cheered the wretchedness of the colony. If it has not utterly perished it survives in men and women who never saw a tree above a stunted alder, or a flower above a chickweed, or any bird save the harsh-throated sea-bird.

From times immemorial the mer de glace has been growing from the central desolation seaward, and pushing out its walls, till now they reach the sea (except along the west up to the seventy-fourth parallel) in tongues that choke up every valley and fiord, and in the great Humboldt Glacier, which faces the sea in an ice-wall sixty miles long and three hundred feet high. Greenland is too small for its ice-robe, which not only reaches the sea but invades it. The Eisblink Glacier plunges deep into the sea and forms a promontory thirteen miles long. The face of the Humboldt has pushed the sea back—how far we have no means of knowing.

The Swiss glaciers, as we saw, creep down from the mountain slopes into the valleys, where they melt and flow on to the sea in rivers. The Greenland glaciers push boldly into the sea and do not melt, but along their entire frontage shed icebergs as a mountain sheds avalanches. The icebergs float southward and melt and discharge their freight in warmer seas. Thus in the seemingly rigid ice-sea the courses of nature are not arrested. There is motion. What is its cause?

Glaciers are made and moved by the sun. Every atom of a mer de glace rose volitant from the face of a sea or lake and sailed up into the higher reaches of air, where it was chilled into a crystal of snow which came down on a mountain or its environing slopes, if the glacier is of our zone—anywhere on the earth's face, if the glacier is in the arctic zone. The evaporating and lifting force is the sunbeam. Experiments which we need not detail, show that every pound of a glacier has cost the sun enough heat to melt five pounds of iron. The mer de glace of Greenland, as we know from the stranded icebergs it has shed, attains a thickness in certain places of three thousand feet and covers an area which we place, approximately, at six hundred and fifty thousand square miles. So vast is Greenland's store of *cold!* Picture this mer de glace as a boiling "mer de fer"— so vast would be Greenland's dower of *heat!* So much heat has nature used in creating so much cold. The heat has been poured out from the sun and expended, little by little, day after day, century after century, on the face of tropical and temperate seas. If so much heat has gone to the making of such an ice-sea, how much heat will be required to move it?

The cause of glacial motion has been one of the most vexing questions in science. Mr. Croll's solution seems entirely satisfactory. It could not have been reached until science in the person of such votaries as Meyer and Joule and others had worked out the law of Correlation and Persistence of Force.

Agassiz's experiments from the hut on the Aar showed that the ice in the heart of a glacier is comparatively warm—warmer in winter, by many degrees, than the atmosphere without.

Faraday showed that when two pieces of ice, moist at the surface, are brought into contact they immediately freeze together.

Tyndall showed that when ice at thirty degrees is made by pressure to change its shape, it is broken into fragments and the fragments instantly reunite and form a solid mass.

Agassiz's borings in the Aar showed that heat can be transmitted through ice, and later experiments have shown that heat can be passed through ice at thirty-two degrees and still the ice remain solid. What becomes of such heat?

Meyer's experiments, and those of Joule and Helmholtz and Grove and Tyndall, have shown that all physical forces are in correlation with heat, and that no force, not even a degree of heat, is lost.

We are ready now for an application of these discoveries in the latest and, we trust, final theory of glacial motion.

Water, in freezing, gives out heat and expands. Ice, in melting, absorbs heat and contracts. Now when the rays of the sun fall on the face of a glacier they cannot raise the temperature of the ice. Molecules on the surface absorb the heat, liquefy, contract, and by the force of gravity move down. In an instant these molecules freeze again and in the act of freezing give up the heat which melted them to melt their neighbors. These, in turn, being liquefied, move down and in freezing again give up the heat which the instant before had melted them, to their neighboring molecules. And thus the sun's heat is carried from molecule to molecule through the entire mass. And thus, while acting as conductors of heat, the molecules change from a crystalline to a fluid condition, and, by the force of gravity, move down from a higher to a lower position. We are to picture these transitions of a molecule from solid to fluid and from fluid back to solid, as almost instantaneous. And the regelation which Faraday and Tyndall showed occurs between sundered masses of ice when brought face to face, we must picture as occuring between the sundered molecules. A wave of heat, pulsing through a glacier, clips molecule from molecule, but the molecules, riven asunder while transmitting the pulse, are welded together the moment the pulse-wave has passed. The melting and moving down of molecule after molecule explains the motion of the glacier. The regelation of the molecules explains the solidity of the glacier.

If the molecules of a mer de glace move, although not simultaneously, and each molecule moves, by the force of gravitation, *downward*, the whole mass of the mer de glace must move, and in the same direction. If the sun's heat were to transform a glacier into water, all at once, the water would flow at once to the sea. Now as the sun transforms into water not the mass but the molecules, and these not permanently but transiently, and not simultaneously but consecutively, and as each molecule moves for an instant as it would move continually if all the molecules were permanently and simultaneously changed into water, the glacier moves slowly down the same gradient that would carry a river.

The sun's heat, falling on rocks, changes their temperature, not their condition; falling on glaciers, it changes their condition, not their temperature. A wonderful economy is this. The rays of the sun, falling on our hills, could not send them into the sea until they had melted them like wax. But the sun-rays that feebly beat on the Greenland mer de glace pulse and throb through the vast piles of ice, snap for a moment the crystalline fetters of every molecule, and send the cold, colossal desolation creeping to the sea.

We go again to Switzerland. Between the valleys of the Rhine and Rhone are the Jura, a range composed of limestone and trending from northeast to southwest. Bounding the "Strath" or Valley of Switzerland on the southeast are the Alps, a range composed of granite and gneiss and having the same trend as the Jura. We follow the Rhone up to its source in the Alps. We find its source, the Rhone Glacier. This glacier, fed by constant accretions of snow along the alpine slopes, moves slowly down into the valley where the heat cuts it off and transforms it from the Rhone Glacier into the Rhone. We stand along its terminal wall and see the piles of rubbish it has brought down. They lie here stranded and take the name of "end moraine." We notice that the Rhone as it issues from the ice is turbid and milky. Taking a cup-full of this water and letting it stand till it settles, we find the sediment it leaves in the cup, a fine impalpable mud. It is rock, ground into atoms by the great ice-mill above. It will be borne away by the river and thrown down, its greater portion, on the bed of Lake Geneva. It dries and presses into a hard sticky clay like the bowlder clay which forms everywhere the base of the glacial drift. We pick up here and there in the rubbish pushed along and piled up by the glacier, polished and scratched pebbles like those which abound in the bowlder-clay or till. If we were standing on a *beach* we could not pick up pebbles like these. Rounded and polished pebbles we would find but not *scratched* ones. Water rolls and tumbles the pebbles on a shore and rounds them and smooths them, but only a glacier scratches them.

From an icy vault in the terminal wall of the glacier we see issuing forth a rapid stream of gray snow-water. We see the birth of what will flow five hundred miles and discharge itself into the Mediterranean as a mighty river — the Rhone, the classic *Rhodanus* "which issued from the gates of eternal night at the foot of the pillars of the sun." We see the birth of fables as well as rivers. To Italian imagination the ice-vault became "the gate of eternal night" and the Gelmerhorn and Gertshorn and Galenstock which tower above, became "the pillars of the sun."

We go now right into the glacier, for here is a grotto hewn hundreds of feet into the ice. The ice overarching us we see is heavily freighted with pebbles, and if we pick into the floor of ice under our feet, we find the ledge over which it creeps, abraded and polished, and, wherever the grinding ice holds a pebble, we find the surface scratched or grooved.

We come out and inquire of old inhabitants whether the glacier remains always the same or whether it ever advances and retreats. They will tell you that within their memory, in 1818, the ice crept down a hundred and fifty feet further into the valley. If we search the ledges within this area we find them abraded and grooved. As we have seen the ledges beneath the glacier *undergoing* the process of abrasion and grooving, we will say that the testimony of the rocks answers to the testimony of the men.

We descend the Rhone Valley eighty miles to its bend at Martigny and find, all along, graven on the ledges the same record we had found under the ice and at its foot. But no man has seen the glacier at Martigny and told of it in history or fable. Human testimony we have left far behind and we have only the testimony of the rocks to tell us that once the glacier crept down this valley as far as Martigny.

From Martigny we follow the bend of the valley till it opens into the Valley of Switzerland, and find the ledges everywhere abraded and polished and furrowed. The glacier did not stop at Martigny but pushed on through the valley toward the lakes and the Jura. And right on we follow its foot prints in ledges rubbed and grooved, and in granite bowlders stranded all over the Strath and up three thousand four hundred feet along the flanks of the limestone Jura. The testimony of men told us that fifty-six years ago the Rhone Glacier crept a hundred and fifty feet further down its valley. The less fallible testimony of the rocks tells us that in some long ago this glacier moved down through the Rhone Valley which it gorged to the brim, and overflowing into the Swiss Valley, pushed on and abutted against the Jura to the height of three thousand four hundred and fifty feet! From this sea of ice the tops of the highest Jura rose as islands of snow, while the lower ridges were overtopped by the advancing flood which dropped its Alpine freight in the Valley of the Rhine. From the wall of the Jura a portion of the flood was deflected to the southwest. Following the southwest trend of the Jura we find the course of this flood recorded in grooved ledges, and in Alpine bowlders strewn over the earth at Geneva and Aars and Sattonay and Bublanne and Lyons, and on still further, at Valence, less than a hundred miles from the mouths of the Rhone in the Gulf of Lyons. What a record is this! The Rhone Glacier and Finster Aarhorn and mer de glace are but the shriveled remnants of a mighty ice-sea from whose crystalline deeps the ridges of the highest Jura peered as reefs over frozen waves, and the domes and needles of the highest Alps rose in islands of eternal snow!

This is not all. Other events are recorded on and in the Swiss till, a clay which Swiss geologists have called "ground moraine." This till has been subjected to great pressure and it carries many striated pebbles. If we imagine the fine clay which issues with a stream from a glacier, and represents the rocks which the ice has abraded and ground into powder, if we imagine this in immense quantities subjected to immense pressure and holding here and there a seam of gravel or sand, we will have, in imagination, the till. Now this till is overlaid in Switzerland by alluvial sand and gravel, and on the alluvial, in the cantons of Zurich and St. Gall, beds of lignite occur. The lignite varies in thickness from two to five feet. It is composed chiefly of peat-moss but it holds the remains of many trees, such as Pines and Oaks and Birches and Larches, and the remains of Elephants and Cave Bears and Stags and the Urus and Rhinoceros, and the wings of insects. Prof. Heer has made a careful study of the fossil plants and they tell him that the climate of Switzerland, in their day, was essentially the same that it is now. The great ice-sea had shrunk back from the Jura and shriveled into such glaciers as the Finster Aarhorn and Rhone. A forest of Oaks and Elms and Birches and Pines grew up from alluvial, spread by water of the melting ice-sea, over its ground moraine. Elephants were roaming through these forests, companioned by the Elephantine Ox and the Wool-haired Rhinoceros and Cave Bear, all extinct. The fetters of a glacial winter were broken — broken for a time, a time of whose duration we have no register save the lignite beds of Zurich and St. Gall.

Overlying the lignite are beds of gravel surmounted by granite bowlders from the Alps. Where lignite does not occur, the alluvial spread over the ground-moraine is overlaid by another ground-moraine. This second ground-moraine, which is the underclay of a second mer de glace, is not commensurate with the first. The end-moraines left by the second ice-flood are still preserved and mark its limits.

The facts and the deductions which must be drawn from them may be presented to the eye as follows:

Fact.—Planed and striated ledges and alpine bowlders, from Rhone glacier down Rhone Valley and over great valley between Alps and Jura, up three thousand four hundred feet on slopes of Jura, over Lower Jura to Rhine and along trend of the Jura southwest to Valence.

Deduction.—A mer de glace of which the Rhone glacier is a shriveled remnant buried all Switzerland and crept over the Jura into Germany and invaded France to the latitude of Valence.

Facts.—Beds of gravel and sand overlying the bowlder-clay; over the gravel, alluvium holding lignite and fossil trees and remains of animals characteristic of warm latitudes.

Deductions.—That a change of climate ensued; that the mer de glace shrunk and, retreating from the Jura, discharged its rubbish of gravel over the plains of Switzerland; that rivers from the melting ice spread a coat of alluvium over the glacial rubbish; that a luxuriant forest-growth, supporting Mammals akin to those of the warm zone, spread over the area which the ice-sea had covered.

Fact.—Another ground-moraine, and other end-moraines overlying the old moraines and their cap of gravel, alluvium, and lignite.

Deductions.—That another change of climate ensued; that the warmth which had nourished the oak and the elephant gave way to arctic cold, and the glaciers, leaving again their mountain-valleys, crept slowly down into the plain and pushed their end-moraines nearly to the Jura.

Intercalated between two periods of arctic cold was a period of warmth and verdure. This is the testimony of the drift deposits of Switzerland.

It would detain us too long and cumber the memory with too many details if we were to sketch the history of the Ice Age in other parts of Europe and in Asia. Take this summation by Geikie, and understand that it rests on evidence the same in kind as that we have marshaled from Switzerland: "All Northern Europe and Northern America disappeared beneath a thick crust of ice and snow. This great sheet of land-ice leveled up the valleys of Britain and stretched across our mountains and hills. Being one connected or confluent series of mighty glaciers, the ice crept ever downward and outward from the mountains and pushed far out to sea, where it terminated at last in deep water, many miles away from what now forms the coast-line of our country. This sea of ice was of such extent that the glaciers of Scandinavia coalesced with those of Scotland on what is now the floor of the North Sea, while a mighty stream of ice, flowing outward from the western sea-board, obliterated the Hebrides and sent its icebergs adrift in the Atlantic.

In like manner massive glaciers, born in the Welsh and Cumbrian Mountains, swept over the lowlands of England and united with the Scotch and Irish ice on the bottom of the Irish Sea. At the same time the Scandinavian mountains shed vast icebergs into the Northern ocean and sent southward a sheet of ice that not only filled up the basin of the Baltic, but overflowed Finland and advanced upon the plains of Northern Germany."

That is a picture of Europe. How shall we picture America? By drawing our picture of Greenland on a larger scale. By answering the question we raised when sketching the memorials of a flood from the shores of the Arctic Sea to the latitude of Philadelphia, and from the level of the sea to the Lake of the Clouds — a flood of which we were making the facts proclaim only the area and the depth, not the kind — by answering, "An ICE-FLOOD."

We want details enough to show us whether the Ice Age in America was continuous or interrupted. We take a few sections from the Geological Reports of Illinois and Ohio, and a few from the sea-board in New England. The drift under Bloomington, Illinois, gives the following section:

1. 10 feet of surface soil.
2. 40 " " blue clay.
3. 60 " " gravelly "hard-pan."
4. 13 " " black mould, with fragments of wood.
5. 89 " " hard-pan.
6. 6 " " black mould.
7. 34 " " blue clay.
8. 2 " " quicksand containing fossils.

We shall find more instruction here than the array of commonplaces would seem to promise. At three horizons, 8, 6, and 4, we find memorials of forest life.

Under Columbus, Ohio, the drift is arranged as shown in the following section:

| | | |
|---|---|---|
| 1. | Surface earth, . . . | 1 foot. |
| 2. | Sand and gravel, . . . | 2 feet. |
| 3. | Blue clay, with bowlders, . . | 4 " |
| 4. | Sand and quicksand, . . . | 5 " |
| 5. | Clay packed with ruins of a buried forest, | 1 " |
| 6. | Blue clay and sand, . . . | 18 " |
| 7. | Clay, sand, and gravel, . . | 12 " |
| 8. | Bowlder-clay, . . . . | 68 " |

Bones of the Mastodon and the great Beaver have been found in the clay of the buried forest. The old forest-bed appears everywhere under large areas of Ohio, and among its trees were the Beech, the Hickory, the Sycamore, and, most abundant of all, the Red Cedar. When we compare this section with others in other parts of the State, and with that in Bloomington, we find it to register a succession of events as follows:

Bowlder-clay, equivalent to ground-moraine of Switzerland, is the memorial of

A vast mer de glace, the first and greatest ice-sea.

Blue clay and sand without bowlders chronicle

A change of climate; mer de glace had retreated and a lake of fresh water covered a good part of the area it had left.

Buried forest and associated Mastodon remains chronicle

The silting up of the lake and appearance of land and trees and the greatest of land Mammals.

Sand and quicksand testify of

Flats, marshes, river-floods. A period of submergence.

Blue clay, with bowlders testifies of

Submergence continued till the land was plunged beneath a sea; Arctic cold coming on again; Icebergs floating southward and dropping their bowlders.

Sand, alluvium, shell-marl, tufa, surface-earth, testify of

Emergence from iceberg-sea and return to warmer climate.

Dr. Newberry's generalization from Ohio sections is: *glacial* drift, lake sediment, forest-bed, *iceberg* drift, alluvium. Expand this statement a little and you have our own readings from the Illinois and Ohio sections. The facts of chief significance is the intercalation of a warm period between two cold periods, the first, a period of intense cold and continental glacier; the second, a period when the cold was less intense and the land was buried, not under an ice-sea but a sea of floating icebergs.

We have made no accout of the mounds and cones and ridges of gravel and sand so common in New England and the Northwest. Such mounds and ridges are often sprinkled over with bowlders, but do not often contain bowlders within. They correspond in form and position and structure to the "kames" of Scotland. Prof. Geikie has figured a bowlder as it lay in the heart of one of these kames, and it is so instructive that we have reproduced his drawing (Fig. 44). You will see how the laminæ bend down under

Fig. 44.

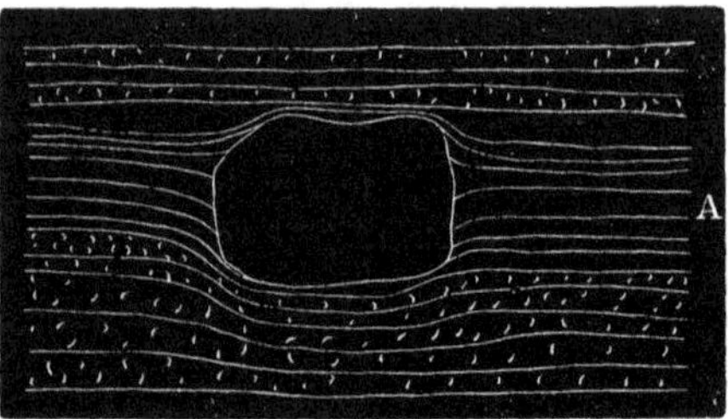


the bowlder. You will see that the thin leaves up to the laminæ marked *A* had been spread out on the bed of a sea or estuary before the bowlder was dropped. When it fell its weight bent down the laminæ and buried half its bulk in their yielding bed. A glacier had pushed out to sea and shed an iceberg, which floated down on the current and dropped from its freight this bowlder. The laminæ from *A* downward antedate the event, from *A* upward they follow it.

These ridges and mounds of sand or gravel, or sand *and* gravel must have been rounded up by currents of water. An occasional bowlder, dropped into them, shows that the currents were those of a sea, and, as the bowlders are rare, a sea with but few icebergs.

Along the Atlantic coast from Long Island to the Arctic Zone are beds of clay and sand which corroborate the testimony of the sandy knolls and the western sections. These beds have been named after their characteristic fossils, the *Leda clay* and

Fig. 44. A bowlder in laminated clay and sand.

*Saxicava sand.* If you imagine a line drawn from Sancati Head, Nantucket, a bluff eighty feet above the sea-level, to a point on Beechy Island in Barrow's Straits, a thousand feet above the sea-level, you will have the gradient on which these post-glacial beds of clay and sand rise from Buzzard's Bay to the Arctic Zone. And of course you will have the gradient of post-glacial submergence. The land in the polar regions sank a thousand feet for each eighty feet it sank in the latitude of Nantucket. Now the clay beds are from deep water and on the coast of Maine they carry an assemblage of fossils which indicate sub-arctic conditions. The overlying Saxicava sand is from shallow water and holds an assemblage of fossils which indicate a warmer sea. We cannot possibly mistake the concurrent testimony of the drift sections, the sandy knolls and ridges, and the Leda clay and Saxicava sand. The Leda clay tells us that after the northern portion of the continent had been buried under a mer de glace which in New England, attained a thickness of more than five thousand feet, it was carried down and buried under an ocean whose temperature in the latitude of Portland, Maine, was sub-arctic. It tells us that the downthrow was vastly greater in the north than in the south. Let us remember now that the records which we translate, in some places are effaced, and in all places are fragmentary. We must piece out the evidence of one stratum or one locality by that of another. The Leda and Saxicava beds tell us only of post-glacial submergence and sub-arctic cold, and emergence and returning warmth. Our western sections supplement this evidence and give us certain missing links in the chain of events which followed the melting of the mer de glace. The clay, (No. 8,) tells that fresh water lakes covered a great part of the area left by the ice. This is precisely what *must* have occurred. In cold latitudes a flood almost always terminates a severe winter and a *great* flood must have terminated the glacial winter. An immense sheet of fresh water of which clay No. 8 is the memorial, covered the area of our northern lakes and spread southward over Ohio and Indiana and Illinois. Clay No. 5 is our witness that when the water was drained off, the twice desolated land, under warmer skies supported a luxuriant forest growth and gigantic beavers and elephants. No memorials of these fresh water floods and inter-glacial period of warmth are preserved in the drift bordering on the Atlantic. No. 3 of the Illinois drift adds now its testimony to that of the Leda clay and tells us of submergence beneath a cold sea — cold as shown by the Leda clay, and choked with floating icebergs as shown by the bowlder-strewn clay of the west. Finally, the knolls and ridges west and east, sculptured into the forms they wear by currents that swept over the bottom — consequently a shallow bottom — add their testimony to that of the Saxicava sand and tell us of the uplifting of the sea-bed and the coming on of a milder climate — a milder climate as shown by the Saxicava sand in its fossils and by the sand-knolls in their comparative freedom from bowlders.

If now we were to push such investigations still further and inquire into the details of the drift in England and Scotland and Ireland and Scandinavia and Germany, as we have into those of Switzerland and America, we would find everywhere the evidence of a warm period intercalated between periods of intense cold, and almost everywhere we would find evidence of a submergence following the retreat of the first and greatest mer de glace. Indeed the correspondence between the drift beds of Europe and those of America is so close that we find, at the same horizon, the equivalent of our clay with buried forest, and higher up, on the same horizon as our Leda clay, its equivalent in the shell-clays of Scotland and Scandinavia.

And now, of movements so vast and over vast areas so uniform, we must search for the cause in no fitful "convulsions." The causes must have been as orderly as the events. And they must have been commensurate with the events. No collocation of land and water, accidental or designed, could have brought about the glacial epoch. No piling up of mountains or polar snow-caps could have tipped the earth's pole and brought it about. No forces, chemical or physical, known to dwell in the earth's core or crust, or known *ever* to have dwelt there, could have brought it about. The causes must have been, not tellurian, but cosmical. "Our position in the solar system and the motions and form and equilibrium of the mass of our world among the planets," says M. Comte, "must be known before we can understand phenomena going on at the surface."

Every school-boy has learned that the earth's path around the sun is not a circle but an ellipse, and that the sun is not in the center of this figure but in one of the foci. This form of the orbit would bring the earth nearer the sun in one part of its journey than in another, but difference in the sun's distance, as the school-boy knows, does not cause our seasons. If you imagine the earth's path in space to be extended into a plane, the axis would not be perpendicular to this plane but tipped at an angle of twenty-three and a half degrees. As the angle of the axis with the plane of the orbit — called the plane of the ecliptic — is nearly constant, at one point in the earth's journey the pole will tip over toward the sun and at the opposite point the same pole will tip *away* from the sun. The point on the ecliptic nearest the sun is called *perihelion*, and the point furthest away is called *aphelion*. At present the north pole tips toward the sun in aphelion and our hemisphere, being flooded with sunshine, enjoys a summer. At the opposite point of the ecliptic the same pole, being turned away from the sun, our hemisphere, although three millions of miles nearer the sun than it was six months before, passes through its winter.

At two other points on the ecliptic the axis is turned at right angles to a line drawn from the sun to the earth, and while we are passing through these points the sun's rays bathe the earth from pole to pole and everywhere the days and nights are equal.

We have said that in our times the north hemisphere enjoys its summer in aphelion, but this state of things has not always prevailed nor will it continue always to prevail. If the earth were an exact sphere July in our hemispere would always be midsummer. But as the earth is pressed in at the poles and swollen out at the equator, the pull of the sun and moon on the equatorial swelling slowly changes the position of the axis. If we imagine the axis prolonged into the heavens we would find that under the influence of that "pull" it would describe a revolution among the stars. The period of that revolution would be twenty-six thousand years, but a movement of the major axis of the orbit, caused by the pull of the other planets on this, reduces the period to twenty-one thousand years. Imagine now a plane through the earth's equator extending outward into the heavens. As the axis describes a revolution in twenty-one thousand years, of course, it carries the plane of the equator with it. Imagine another plane, a plane from the sun perpendicular to the axis of the earth. Such a plane would cut the orbit at two points, the points of equal day and night. Now as the axis performs a revolution in twenty-one thousand years, carrying with it the plane of the equator, the points at which these two planes cut each other must be carried along with the axis and complete a revolution on the ecliptic in the same time as the axis describes a circle on the heavens. These equinoctial points move back along the ecliptic in one year the same distance which the earth travels in twenty-two minutes and twenty-two seconds. Let us fix our attention on the autumnal equinox. We cross this point on the ecliptic now at a certain hour on the twenty-second of September. Next year we will reach it twenty-two minutes sooner. In sixty-five years, (omitting fractions) we will reach it a day sooner. In five thousand four hundred and sixty years we will reach it eighty-four days sooner and our autumnal equinox will fall on the first of July. In ten thousand five hundred years (making account of the twenty seconds) we will pass the autumnal equinox on the first of March and then March will correspond in season to our present September and the mid-winter months will be July and August. Those who live in our hemisphere ten thousand five hundred years hence will have short and hot summers, and long cold winters. In the south hemispere whose seasons occur now as ours will occur then, glaciers in Patagonia creep down to the sea-level on the same parallel as Paris, and in New Zealand a glacier reaches the sea in the same latitude as Portland, Maine.

Such revolutions bring, alternately, the snow-cap of the North and South Poles down near the cold-temperate zones, but they are not an adequate cause of glacial epochs.

The earth's orbit is as inconstant as its axis. At times there is such a collocation of the planets as to cause their united pull on the earth, at certain points of its journey, to draw it away from the sun and so to increase the ellipticity of its orbit. At other times their collocation is such as to cause their pull to diminish the ellipticity. The periods of maximum and minimum eccentricity occur at very irregular intervals, and to calculate them has required the highest skill of some of the highest mathematicians of the age. Science is especially indebted to Mr. Croll, who, at the cost of immense labor and patience, has calculated the form of the orbit as it has been for a million years past, and as it will be through a million years to come. From his tables as a basis we have constructed a chart which will picture to the eye the leading changes which have occurred through the past million, and those which are to occur through the coming million years. A line drawn from one of the foci to what would be the center, if the figure, without enlarging the periphery, were changed into a circle, is a measure of the ellipticity. At present the length of such an imaginary line is three million miles. Let us suppose a number of circles with their centers along the line *a b*. Let us suppose, now, that these circles are drawn out into ellipses. The line raised from *a b* at *o* expresses the form of the orbit as it is *now*. Its length would be three million miles, but we designate it 3, as we shall designate all the lines by the figure which expresses the number of millions or the

Fig. 45.

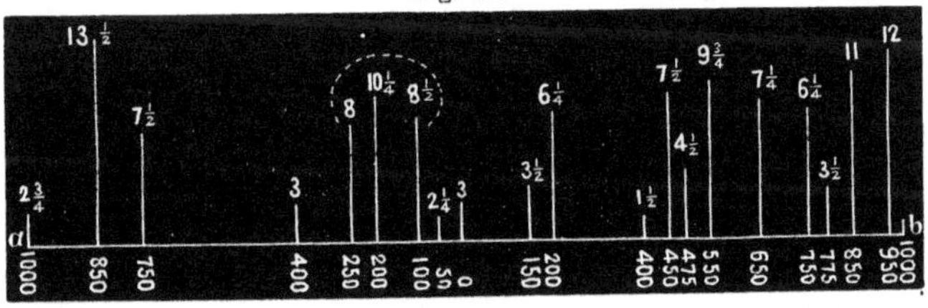

fraction of a million. The figures along the base-line shall designate the years by the number of thousands —from *o* to *a* the years past, from *o* to *b* the years to come. The line raised at 50 is 2¼, and shows that fifty thousand years ago the orbit was more circular than now. The line from 100 is 8½, and shows that one hundred thousand years ago the orbit was far more elliptical than now, the ellipticity having reached eight million five hundred thousand miles. The line from 200 is 10¼, from 250 it is 8, and from 400 it falls back to 3. From 750 it rises to 7½, and from 850 it rises as high as 13½. At 1,000 it has fallen to 2¾, showing that one million years ago was a period of minimum eccentricity. We look, now, from *o* along the right, and see the line from 150 standing at 3½, showing

Fig. 45. Time-scale of Glacial Epochs.

that for one hundred and fifty thousand years the eccentricity will remain almost constant. At 200 the line rises to 6¼. At 400 it falls almost to the base-line, being only 1½. At 450 it has risen to 7½, and at 475 it has fallen to 4½. At 550 it stands at 9¾; 650 at 8¼; 750 at 6¼; 775 at 3½; 850 at 11; 950 at 12, and 1,000 at 1, showing that one million years from now the form of the orbit will approach very near to a circle.

Looking back beyond the zero of our line, we see that a period beginning about two hundred and fifty, and closing about one hundred thousand years ago, was one of great eccentricity. Two hundred thousand years ago this period culminated in an eccentricity of ten and a half millions, and the earth in aphelion was ten and a half million miles further from the sun than in perihelion. For a hundred and fifty thousand years the earth's aphelion distance from the sun was more than eight million miles greater than its perihelion distance. What effect would this period of great eccentricity have had on the climate?

Mr. Croll has shown that if the winters of our hemisphere should occur in aphelion during a period of maximum eccentricity, we would receive nearly one-fifth less heat during winter than we receive now, and in summer nearly one-fifth more. The winters would be about thirty days longer than they are now. But as in summer we would be correspondingly nearer the sun, and would receive an excess of heat corresponding to the loss of heat in winter, would not the short and hot summers melt away the accumulations of the long and cold winters? Mr. Croll has shown that they would not. We are to remember that the direct rays of the sun pass through the air without heating it. In Northern Greenland, where the rays of a midnight sun pour down on the earth so warm as to start the tar on a ship, yet in spite of such heat in the sun-rays, and bombardment by such rays from an unsetting sun through six months, the summer temperature remains about the freezing point. The air is warmed by *radiant* heat, heat which has first struck the ground and has been thrown back by the sand or the naked rocks. Before the summer heat which falls on Greenland can give her a summer they must melt away her mer de glace and rebound into her atmosphere from the ground-moraine and sculptured rock-ridges. But the heat is expended in pulsations that throb from molecule to molecule of the ice-sea and send its vast bulk creeping slowly to the ocean. Now our hemisphere is three million miles nearer the sun in winter than in summer. Let the earth sweep away on her path ten million miles *further* from the sun during our winter, and let the winter be thirty days longer than now — the hot sun-rays that would beat on our hemisphere during the short summer would melt away enough snow and ice to load the air with vapor, but not enough to lay bare the ground and break all the shackles of the arctic winter. A glacial epoch would ensue. The climate of Greenland would spread southward to the borders of Maryland.

Let us take, now, this period of great eccentricity, one hundred and fifty thousand years, and divide it by the period of equinoctial revolution, twenty-one thousand years, and we will see that seven times during the "Glacial epoch" the winters of our hemisphere must have occurred in aphelion, and seven times those of the South hemisphere. Neither hemisphere could have remained in glacial fetters more than ten thousand years at a time, as the revolution of the equinoxes would, in ten thousand years, carry its winter into perihelion, and perihelion winters, occurring so near the sun and being so short, must have been warm and genial.

We understand, now, why we should find, intercalated between the memorials of cold epochs, the record of periods of warmth and life. If our American drift records only one such period, the Scotch records more than one — and all records, remember, are but fragmentary.

And a little reflection will make us understand, too, why, above the memorials of the great ice-seas, we should find the evidence of widespread submergence, submergence deepest where ice was thickest. Lyell has noted the fact that solid rocks like granite and sandstone expand and contract annually, even under such a moderate temperature as that of a Canadian winter and summer, and Deville has shown that granite may contract, in cooling, at least ten per cent. If the crust shrinks perceptibly under the chill of a Canadian winter, how much would it have shrunk when it lay, from the Arctic Sea to the thirty-ninth parallel, and, perchance, for ten thousand years, under the embrace of a continuous and colossal mer de glace?

We turn again to the chart and glance from *o* along the segment of the line which leads into the future.

For one hundred and fifty thousand years the climates will suffer no appreciable change from the inconstancy of the orbit. In two hundred thousand years the eccentricity will be double what it now is and the winters will be decidedly colder than now. Slighter oscillations will continue to occur till somewhere about the year of grace 851,875, when the orbit will suddenly swell out into an eccentricity of eleven millions, which, about the year 951,000, will increase to twelve millions. If we can rest in this latest theory of Glacial Epochs — and science has no other — we may set our house in order and await the coming of a great ice-flood toward the close of the nine hundredth millenium. As the equinoxes will complete nearly five revolutions within the period of the coming cold, each hemisphere will in turn be glaciated.

We glance now from *o* along the segment which led our minds back into the past. Within the past million years there have been three periods of great eccentricity. Suppose the line projected into the *infinite* past. Ever since our solar system was born and cosmos came out of chaos, the sister planets have been pulling on the earth and changing her orbit. Here and there along the projected line a perpendicular would rise to tell of a period of great eccentricity, and if the record were not effaced, the rocks synchronous with the perpendicular would tell us of a Glacial Epoch.

# CHAPTER VI.

**Man's Antiquity — His Migrations — His Development from Savagery to Civilization — The Historian Baffled — Science can See where History cannot — The Pre-historic Chinaman — Pre-historic Indian — Pre-historic Dravida — Pre-historic Irishman — Man in Asia, Africa, Europe, America, and the Pacific Islands on the confines of Brutedom, with no Laws, no Institutions, and no Memory of a Past — Shell-heaps — How long since the Indian Possessed America — Mounds and the Mound-builders — Man was in California before the Mound-builder was in the Mississippi Valley — Paleolithic Age — Neolithic Age — Bronze Age — Iron Age — The Paleolithic Man Restored — His Tools — His Abodes — His Destruction — Emergence of Neolithic Man.**

THE sciences are interdependent. Geology learns from astronomy the cause and date of a Glacial Epoch and teaches to history the first chapter in the life of man. It was written in the age of ice.

We revert with strange fascination to the period of infancy. America at her centennial and Iceland at her millenial bring from cellar and garret and chest and crypt, heirlooms to keep green the memory of the nation's childhood. From dark crypts in the earth science is bringing heirlooms that create in the mind a picture of the *race's* childhood.

The past holds us as by a spell. We look back through the pages of history till we find names and dates and boundaries fading out in a mist of fable. We look beyond the horizon line where history melts into legend and discern only the flitting to and fro of forms, dim and shadowy. In vain we strain our vision to find something that will give us *parallax*, some fixed Tadmor in the wilderness of wanderings. And in vain we hear the cry of despair which comes to us from a baffled historian: "We *must* give it up, that speechless past; whether in Europe, Asia, Africa, or America; at Thebes or Palenque, on Lycian shore or Salisbury Plain, lost is lost; gone is gone forever."

In vain this cry, for we will *not* despair. Baffled on one line of approach, with better equipments we attack the problem on another. What though Bunsen and Lepsius assure us that a Pharaoh sat on the throne of Egypt five thousand seven hundred years ago, and that beyond this date is a wall of darkness even like the fabled darkness which invested Egypt at the stretching forth of a hand toward heaven? *We* stretch forth our hands toward the *earth* and pluck from the Nile-mud which underlies the Sphinx and the Pyramids, a flint arrow-head or knife whose fashioning may antedate the Sphinx by as many centuries as the Sphinx antedates the Roman Colosseum. And it lifts the curtain from Egypt's night. It tells us of a time when the Egyptian had not learned the use of metals and when he wrought of stone, weapons wherewith to satisfy the craving of the stomach and had no mind to brood over the mysteries of being he came in after ages to embody in the Sphinx.

And what though Remusat tells us that four thousand two hundred years ago, in the morning twilight of history, an Emperor sat on the throne of China, and that beyond the twilight is no light at all? Where history cannot see science may see. China, who sits within her walls as the Sphinx on her sand,

and bides the flow of the centuries as immovable as she — China is fossil and belongs more to geology than to history. Dealing with the mind of China as with any other fossil, we are able to call up and picture to our minds the Ah Sin whose sins and customs hardened and petrified into the sins of to-day.

Surnames in China are fossil tribe-names. The ancestral Fus were the Fu tribe, and the ancestral Sins were members of the Sin tribe. Now at a certain stage in the development of a tribe it comes about, in the struggle for life, that men of one tribe capture their wives from another tribe. This is the Sabine marriage. The Romans lapsed into it once, and the Jews more than once. Now custom, in savage life, becomes a chain which a man breaks at the peril of death. Long after the necessity which led to the Sabine marriage has passed, the tribe will tolerate no union of its own men with its own women, but compel each man by force of custom to wed outside of his tribe. A Seneca couldn't marry a Seneca, nor a Huron a Huron. In China one Sin can't marry another Sin, and one Mee can't take to himself another Mee. What is the meaning of this interdiction? Only one interpretation is possible. It is a tribe custom, hardened and petrified into law. It antedates the Empire and has come down from a time when the Chinese were savage, wandering, wife-catching tribes.

China, as a fossil, is more instructive than the fossil birds of Kansas which carry in their reptilian teeth and fishy backbones, reminiscences of their ancestors. A Chinaman designates his paternal uncle by the same name as his father, a reminiscence of early tribe-times when brothers held their wives in common. He calls his brother's child *his* child of the Chih class, the first name implying that the brothers had their wives in common, and the adjunct, being an after-growth, implying that they do not now hold their wives in common. He calls his mother *Mo*, and his father's sister he calls *Mo-ku*, the adjunct being of later date. Strip from the Kansas bird its plumage and you leave it very much of a reptile. Strip from the Chinaman the after-growths in his system of kinship and let his domestic life be what his names of kinship would imply, and you leave him very much of a brute. His sister would be his wife and his brother would share the incestuous couch. Such a man was the primeval Chinaman, although history tells it not, and legends even are dumb. Fossils in the mind and speech have declared it.

This method, applied to other peoples, will yield like results. The American Indian has no history. How old is his race, whence it came, what it was before the pen of history began to record it, are questions which the Indian can answer only as his brother, the Beaver, could answer for his race — answer from whatever of the past survives in him. There is not in the language of any known tribe, in either America, a word which designates the relationship of paternal uncle. A father's brother is called by the same name as a father. But there is not a tribe in which the relationship is not discriminated in *thought*, although it be not in language. There is not a tribe in which a woman is known as the wife in common of two or more brothers, but when the Indian dialects were forming the fact must have answered to the name, and domestic life must have been what the surviving nomenclature of kinship implies. When the Indian first came under the eye of history he had outgrown and cast off a form of polyandria which we find to-day only in the Nilgherry Hills of India and in the wilds of Thibet.

In South India there are about thirty million people who sustain the same relation to the Hindoo that the American Indian sustains to us. They are children of the soil. Now the *India* Indian has the same way of naming his kindred as the American Indian. His nomenclature implies a polyandria which he does not practice, save in the Nilgherry Hills, and there only as a religious rite. Polyandria could arise only in a very low state of tribe-life when the struggle for life was sharp, infanticide common, and women scarce. It might have arisen independently in India and America, but the correspondence between the two systems is so close, even in insignificant details, as to suggest a common ancestry for the two races. Columbus was wiser, perhaps, than he knew when he called the American an *Indian.*

We turn to a map of the globe, and cast our eyes from the Whitsunday Islands eastward along the Society, the Samoa, the Tonga, the Feejee, New Zealand, Australia, New Guinea, Java, over the Indian Ocean through Madagascar to South Africa; and again from the Sandwich Islands eastward along the Marshall, the Marianne, the Caroline, the Philippine, the Celebes, Borneo, Sumatra, and thence again through Madagascar to South Africa. Our eyes have traversed an area which would lie, all but New Zealand and a portion of Australia, within a triangle whose apex is South Africa, and whose base, distant by half the circumference or the globe, extends from the tropic of Capricorn to the tropic of Cancer. This is the ocean-world of humanity, and over it broods the darkness of a starless night. Man hovers on the confines of brutedom, and has no laws, no history, no memory of the past. What was his past? A people whose customs do not tally with its language has either risen or lapsed. When you put your handkerchief in your pocket, and not on your head, you show that you have outgrown the custom of your fathers, who named this indispensable a *kerchief* because, with them, it was a covering for the head. Now over all that island-world man calls an uncle, *father*, an aunt, *mother*, and a brother's or a sister's child, *his* child. But low as he is, his domestic life does not tally with this nomenclature. Below this deep of barbarism has been a lower deep, and these people have had a history, although they remember it not. Its history has been the first and unwritten chapter of *universal* history. When we pruned from the language of the Chinaman's domestic life all the aftergrowths, all the accretions gathered by centuries of semi-civilization, it answered to the barren deformity of the nomenclature of these pelagic people. If their domestic life began in promiscuity, so did that of the Chinaman. Now the domestic life implied in the language of this race would pertain to an earlier and lower condition of the tribe than that implied in the speech of the Indian of America and non-Hindoo of India. It antedates the Sabine marriage. It is the outgrowth of a condition in which infanticide was not common, a condition in which women were equal in number to men, and pairing might go on within the tribe. Both polyandry and polygyny might prevail. That they did prevail in the Sandwich Islands we infer from a custom which missionaries have described, and which is the survival of a domestic state bordering on promiscuity.

The pruned nomenclature, of the Chinese and uncumbered nomenclature of Oceanic peoples express an early condition of humanity, not in Hawaii alone, but the world over. Why did Abraham explain to Abimelech that Sarah, his wife, was his sister through the same *father*, not the same mother? Why do the Irish go through the mock performance of capturing a wife? Why did Herod marry his sister and Cleopatra her brother? Why did the Incas always marry their sisters? The words of the patriarch are as the outcrop of a stratum. They indicate a social condition below that of the patriarch and before the times of the patriarch, a condition in which relationship was known through the mother, as the father was uncertain — a status of which Rome preserved a reminiscence in her legal maxim, *partus sequiter ventrum*, and America, when her slave code translated the maxim: "*the chain follows the mother.*" The wife-catching ceremony of the Irish is a shadow cast down the slope of unrecorded centuries from a reality. The marriage of Cleopatra with her brother, and the Peruvian Incas with Coyas, their sisters, was something more than a shadow of the past. Royalty gathers up the cast-off robes of the past and folds them about herself. The court-dress of to-day was the common dress of days gone by. She conserves the customs and manners of the past as well as the costumes, and incest was regal after it had ceased to be common.

Science, then, will not echo the despair of the historian. Lost is *not* lost, gone is *not* forever gone; and "whether in Europe, Asia, Africa, or America; in Thebes or Palenque, on Lycian Shore or Salisbury Plains," or whether in the solitudes of Oregon or on the last-found island of the Pacific, wherever we find man we find him under the shadow of a mighty past whose echoes faintly reverberate through his manners and speech. And whether in shell-heaps, or mounds, or gravel-beds, or caves, wherever we find heaped-up mouth-refuse, or tools or tombs of men, which time, by after-changes wrought in the earth's surface or life, has placed in the geologic calendar, we find the memorials of a past, "speechless," "lost," "gone forever" but for the crowning achievements of geology.

In all times, but chiefly in early times, that which determines the habitat of the animal has determined the dwelling-place of man. It is the demand of the stomach, not the pleasure of the eyes. And when man was still the serf of nature, and had not subdued her forests or learned to cope with her wild beasts, when he was neither a farmer nor a hunter, his habitat must have been the shores of oceans and lakes and the banks of rivers. He was fed from the water before he conquered a support from the land. Oysters, clams, mussels may be had by the mere putting forth of the hand, and the heaped-up shells, over which time has but little power, mark the feeding-places of primitive men. Shell-heaps are found on the Atlantic coast at Terra del Fuego, in Brazil, on the Caribbean, in Central America, and from the Gulf of Mexico to the Bay of Fundy; along the Pacific from California to Behring's Sea; on the shores of Denmark, France,

England, Ireland, Scotland; on the coast of Tasmania and Australia; on the Malay Peninsula and the Andaman Islands, and along the rivers of Brazil, North America, and Europe. On the coast of Maine, where oysters do not now thrive, we have seen oyster-shells piled into hillocks sixty feet high over acres of ground —a vast monument to the human mouth! For if we overlook the little fragments of crude pottery and the flint arrow-heads which diligent search discovers in these hillocks, we would have to say that some race or tribe has eaten and died and left no evidence that they were anything but *mouths.*

When Florida came under the eye of history it was peopled by Indians who lived by hunting and agriculture. These people knew nothing of the shell-mounds, so common and so large, along the sea and lakes and rivers. Prof. Wyman who searched these mounds, lends his great authority to the statement that they were left by a race entirely different from the Florida Indians. In the oldest mounds no pottery was found. They contain but few implements, and these were wrought chiefly of shell and bone. Bones of animals killed for food are rare, and the indications are that the people who made these shell-heaps had not learned the arts of agriculture or the arts of the chase, but lived almost exclusively on fish and shell-fish. The mounds do not give a very clear account of their age, but some of them are overgrown with trees which were saplings when Columbus set sail for America.

The element of *time* is more clearly brought out in the shell-mounds of Denmark. These mounds occur here and there along the shores of the Baltic, and wear such a garb of antiquity that for a long time they were supposed to be raised sea-beaches. But a Danish naturalist observed one day that all the shells were full-grown or nearly full-grown, while shells of a raised beach, as a moment's reflection will show, would be of all ages and sizes. He observed, furthermore, that species were lying side by side in these heaps which do not live together, and therefore, in a natural deposit, would not be found *dead* together. A thorough exploration was made which brought out the fact that these refuse-heaps are the sites of ancient villages. The villagers lived on the shore and fed on shell-fish. They allowed the refuse to accumulate around their huts or tents, and while time has obliterated every trace of their dwellings and every trace of themselves, his tooth has spared their heaps of mouth-refuse. How long? The shell-heaps do not occur on the shores of the German Ocean, but only on the shores of the Baltic. Now it is known that the German Ocean is pushing in on Denmark, while Denmark is pushing out on the Baltic. As most of the mounds occur very near the Baltic shore, we infer that the villagers reared their huts right along the sea, within easy reach of their supplies. As a few mounds occur back eight miles from the coast, and none along the coast of the German Ocean, we may infer that these people perished so long ago that the waves of the German Ocean have had time to encroach on the land and eat up the site of their huts, and that the land has had time to push out, in places, eight miles into the Baltic.

And the Baltic has had time to change from a sea into a brackish lake. The evidence is of the simplest kind. Oyster-shells are common in these heaps, but the oyster cannot live to-day in the brackish Baltic, except where northwesterly gales drive the salt waters of the Skagger Rack into the Categat. The mussels, cockles, and periwinkles which live to-day in the Baltic are dwarfed, the water being too fresh to allow their full growth, but those found in the mounds are as large as the same shells when found in the open sea. Jutland was an archipelago, and the Baltic received the full flow of the ocean, when a people with little bodies and round heads* and shaggy, overarching brows, a people cultivating no plant and having no domestic animal but the dog, and no tool but a wooden lance tipped with flint, and flint knife and hatchet, built their huts or spread their tents of skin on the bleak shores of this northern sea. Would you lift the veil which the centuries have woven, and look on their domestic life? Here is the picture. It is a picture of the Fuegian, whose wretched life is writing on the shores of the Pacific, at almost the same parallel, a record which will read in future times as this record on the Baltic reads to-day. It was drawn by the greatest observer of the times.

We mean Charles Darwin. "Living chiefly upon shell-fish the Fuegians are obliged constantly to change their place of abode, but they return at intervals to the same spot, as is evident from the pile of old shells. These heaps can be distinguished at a long distance by the bright green color of certain plants which grow on them. . . . . Their wigwams, in size and form, resemble a haycock. Broken branches are stuck in the ground and imperfectly thatched on one side with

* Evidence: The skulls taken from Danish tumuli.

tufts of grass and rushes. . . . . We pulled along side a canoe with six Fuegians. They were the most abject and miserable creatures I ever beheld. They were quite naked. It was raining heavily, and the fresh water, together with the spray, trickled down their bodies. They were stunted in their growth, their hideous faces bedaubed with white paint, their skins filthy and greasy, their hair entangled, their voices discordant, their gestures violent. . . . . At night five or six of these wretched beings, naked, and scarcely protected from the wind and rain of this tempestuous climate, sleep together on the wet ground, coiled up like animals. Whenever it is low water they rise to pick shell-fish from the rocks; and the women, summer and winter, either dive to collect sea eggs, or sit patiently in their canoes, and with a baited hair-line, jerk out small fish."

Here and there, as on the Tennessee River and the American Bottom in Illinois, burial mounds occur in proximity to the shell mounds, as in Denmark tumuli are found near shell-heaps. Some of our fresh water shell-heaps, we can hardly doubt, were the work of the Mound-builders. And who were the Mound-builders?

Over the steppes of Asia from Russia to the Pacific, over all Europe from the Atlantic to the Ural Mountains, and over America from the polar zone to the western slope of the Alleghenies and thence southward to the Gulf, the face of the earth is dotted with mounds raised over the pre-historic dead. In England the mound reached its fullest development in Silbury Hill, a hundred and seventy feet high; in America, at Cahokia, Illinois, it attained its greatest bulk in a pile which covered an area of six acres and rose to the height of ninety feet, while in Egypt the idea reached its culmination in the Pyramid of Cheops. The people who piled up these mounds in America were not the modern Indians. They seem to have been not one race, but two, at the least. Squier and Davis have shown that the ruins of a system of earth-works can be traced from the head-waters of the Alleghany and Susquehanna, through central New York and northern Ohio to the Wabash. But no enclosers, no "battle mounds," are found in the Valley of the Mississippi where the mound-system reached its fullest development. The people along the skirts of the Alleghany were warlike, in the valley of the great river they were peaceful. Along the mountain frontier the greatest works are battle-works. Along the Mississippi, and over most of its water-shed, the only works are those of peace. It is a fact worthy of note that no war-implements are found in these mounds, although other implements are so abundant that we can almost reconstruct from them the Mound-builder's domestic life.

On a bluff overlooking Grand River, in Michigan, in company with a number of gentlemen, we opened a large burial mound. At the depth of six feet we took out a little copper hatchet and a few copper awls. The copper must have been taken from the water-shed between Lake Superior and Lake Michigan. Now the Indian when first discovered by the white man knew nothing about the uses of copper. He called it "big medicine." He had no traditions about the use of copper by his ancestors, and no knowledge of any locality where it could be obtained. It is evident, then, that we were exploring one of the graves of a race which preceded the Indian. If we go now to Ontonagon we shall find the most abundant evidence that copper was mined in this region ages ago. We shall find trenches and pits, heaps of rubble, stone hammers for pounding off bits of copper from the vein and vessels for bailing water from the mine; and all these under conditions which indicate the lapse of many years. Mr. Knapp, the superintendent of a mining company, crept into a cavern along the southern escarpment of a hill, and finding evidence that the opening was an artificial one, he obtained the assistance of three men and explored it. They cleared out the rubbish, and on the floor of the excavation they discovered a vein of copper with ragged projections. Some projections from this vein had been pounded off, and there, in the excavation, were stone hammers which in the hands of an extinct race of miners, had done this very work. They penetrated another excavation, and at the depth of eighteen feet they struck a mass of native copper ten feet long, three feet wide, and nearly two feet thick. They dug around it and found that it rested on billets of oak. Time had taken from the wood all its fiber and the knife-blade passed through it as through muck. The block of copper, which weighed more than six tuns, had been lifted by the ancient miners about five feet and then dropped, being too heavy for their skill in engineering.

The ancient mines of Ontonagon will account for the copper implements in the mound.

At the same level as the copper we took out large plates of mica. Mica is as common in the mounds of the west as copper and as foreign to her rocks as the metal. Where did the Mound-builder get his mica?

This mineral is often found in large plates. Plates have been taken from a Chillicothe mound more than a foot in diameter. Now Prof. Kerr, State Geologist of North Carolina, while exploring some ancient pits in an unfrequented mountain accessible only by two days' journey on horseback after leaving the highways of travel, struck veins of mica. A little examination showed him that the pits were artificial and that he had not discovered but *re*-discovered a mica-mine. The workings showed marks of age similar to the copper-workings of Ontonagon. And as the Ontonagon mines have been re-discovered to supply civilization with its copper, so the North Carolina mines, so lately re-discovered, already supply commerce with the choicest mica. History, they say, repeats herself and moves in a circle. Does she round out her circle by repeating times pre-historic? Circleville, Chillicothe, Newark, Cincinnatti, New Albany, St. Louis, Chicago, Rockford, Rock Island, Dubuque, Davenport, Milwaukee, Madison, Beloit, Detroit, Grand Rapids, Kalamazoo, and almost every other town on the water shed of the Ohio, the Mississippi and the Lakes, are founded on the ruins of Mound-builder towns, and men now traffic in copper from Lake Superior and mica from North Carolina as did the men through whose graves they have run their streets.

Associated with the copper and mica were sea-shells from the coast of Florida. In some of the mounds we find implements made of obsidian from the mountains of Mexico. The Mound-builder had a commerce which extended from the shores of Lake Superior to the mountains of North Carolina, to the coast of Florida, and to the table lands of Mexico.

Pipes and pottery were found. Tobacco was indigenous to America and the ancient American had learned its properties, and *liked* them.

We struck at last what remained of the man for whose ghost these trinkets were intended. It was a portion of the skull. All the animal matter was gone, and what was left was a dark, spongy mass, which was

Fig. 46.

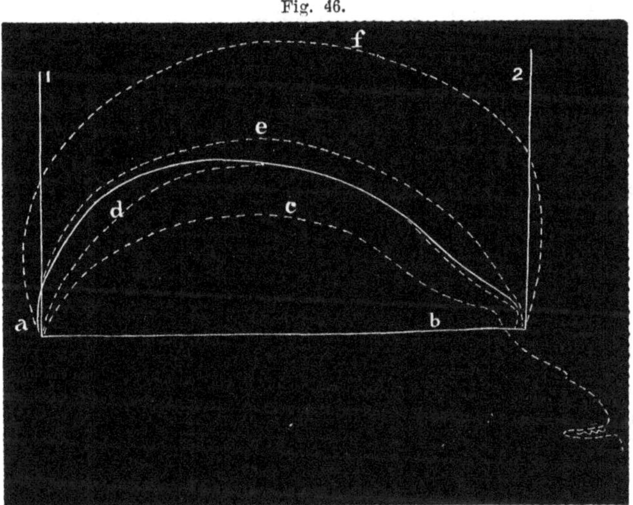


ready to crumble at a touch. The skull is so remarkable that we give a drawing in Fig. 46. The unbroken line represents the Mound-builder's skull; the dotted line, *d*, the Neanderthal skull; *e*, the Australian; *f*, the European; *c*, the Chimpanzee. This Mound-builder had a thick bony ridge over the eye and a low, retreating forehead. These features, especially the thick

Fig. 46. CONTOUR OF SKULLS.—The unbroken line that of the Spoonville Mound-builder; *d*, Neanderthal Man; *e*, an Australian; *f*, a European; *c*, a Chimpanzee; *a b*, Glabelo-occipital line.

superciliary ridges and low frontal, place the man very low on the human scale. Look now at the Neanderthal skull, and you will see that the forehead falls down a little lower, and the superciliary ridge is so prominent as to give to the skull an animal cast. The low forehead intensifies the animal cast, while to the anatomical eye the *backhead* seems as under-human as the forehead. The Neanderthal is the lowest human skull which has come under the eye of science.

The series represented in the plate will take us from the Chimpanzee to the European. The skulls are so placed that a line joining the occipital crest *a*, and the glabella *b*, is horizontal. The superciliary ridge of the Chimpanzee is so prominent as to tell the eye at once that it looks on the cranium of a brute. The low, forward sloping occiput is another animal character. The Neanderthal is not so prominent in the superciliary ridges, the frontal bone does not fall back so abruptly, or the occipital forward so abruptly. Our Mound-builder does not have such a brutal development of the superciliary ridge, but his forehead is quite as bestial. From the glabello-occipital line to the vertex the Neanderthal measures three inches and two-fifths, and the Mound-builder three inches and three-eighths. The backhead of the Mound-man is better than that of the Cave-man, but still no part of the occiput touches the perpendicular drawn from the occipital protuberance, *a*, to the line *a b*. The Australian is perhaps the lowest of living races. The occiput falls back a little beyond the perpendicular, the frontal does not shelve backward so much as in the skull of either of the extinct races, and the superciliaries are not developed. See now the outline of the European skull which crowns the series. See how the occiput swells out beyond the perpendicular and how the frontal is rounded up into a noble arch.

Look now at the Mound-man as if you were looking him in the face. The parietal, or "wall bones" of the skull do not round up, dome-like, as they do in us. They are flattened, so that the head, seen from front or back, tapers up like a pyramid. A low, pyramidal head, a forehead that falls back like that of a brute, great ridges of bone jutting like penthouses over the eyes — with such features *above* the eye, what would be the face-features below?

The frontal bone determines the face. Note the *backward* slope of this bone in Chimpanzee and *forward* slope of the face. In the fish the frontal, falling right back, gives no forehead at all, and the fish has no face at all, only a mouth. In the highest races of men the frontal is upright and the facial parts do not slant forward but are carried downward. In the Mound-builder the backward slant of the frontal must have given a forward slant to the face. The nasal spine sticks out, as Col. Foster has phrased it, like the beak of a bird, and the jaw followed the slant of the nose. The Mound-builder was prognathous. *This* Mound-builder could have been no other than a weak, miserable, bestial, savage. His skull is typical, although it is one of the lowest, perhaps the very lowest which the mounds have yielded.

How could a race with such heads have mined the Carolina mica and the Ontonagon copper? How could a people that could not dig a well pile up at Cahokia a mound one-fourth as large as the Pyramid of Cheops? Peruvian civilization rested on heads smaller than those of any Indian tribe known to-day in either America. Although the Peruvian was weak he wrought mighty works. A weak, stupid, submissive people, when controlled by such chiefs as the Incas or Montezumas could achieve all that was done by the Mound-builder, the Mexican, or the Peruvian. Indeed they were essentially the same people. Leaving out the Eskimo there was but one race in America, North and South. It was that of the Red Indian. The Mound-builder although very different from the average Indian living now, bore a strong resemblance, in the shape of his head, to the Pueblo Indians of New Mexico.*

How long since the Mound-builder perished it is difficult to say. Rivers have had time to encroach

* At the Detroit meeting of the American Association for the Advancement of Science, Dr. Farquharson called attention to certain manifestations of disease among the Mound-builders, the most interesting of which was a peculiar roughening of the articular surfaces of the cervical vertebræ and a bony anchylosis of the joints. In mounds at Rock River and Albany, Illinois, diseased cervicals have been found, which were more or less anchylosed. In one case the second, third, and fourth vetebræ were firmly united and the articular surface of the fourth showed that it had been anchylosed with the fifth.

Disease of the spinal column affecting the dorsal and lumbar region is not uncommon among men or apes, but disease in the cervical region is uncommon. In one form of disease cure is affected only by anchylosis of the joints, and this is possible only when the patient remains in absolute repose for at least one year. Nelaton has been able to note only twenty-five cases of such cure. The inference Dr. Farquharson draws from the cases yielded by the Mounds of Rock River and Albany, is that the Mound-builders were so far advanced in civilization as to house and feed and nurse the sick and keep them in a state of absolute repose, if need be, for a year! To our mind the facts hardly justify the inference. Anchylosis, more or less complete, of the cervical vertebræ is not a very rare occurrence and it is found on the terraces and eat into the mounds they support, and then eat back a retreat of nearly a mile; trees have had time to register eight hundred years of growth in their trunks, and many generations of trees have had time to grow and die and cap the mound with the mould of their decay; and the mounds had time to lose their sacred character and be cultivated in "garden beds" before they were overspread by forests.*

The Mound-builder may have been the first man on the continent, but neither his mound nor his shell heap is the earliest record of man.

The Sierra Nevadas of California have been the seats of great glaciers which descended along their slopes, ground into pebbles and grains of sand their gold-

in our dissecting rooms in subjects who have not been helpless and still.

Paschal used to say that half the ills of humanity result from our inability to sit still in a room. One of the last virtues acquired by humanity is that of *keeping still.* The Indian has it not. We cannot think that the Mound-builder had it.

* In Wisconsin, Lapham's Antiquities.

Dr. Abbot has made investigations which throw much light on the antiquity of the Indian and his growth in the arts. In a forest of beech and oak and chestnut, the increase to the depth of the soil from the decay of leaves is estimated at 1-128th of an inch a year. Now arrow-heads of jasper and quarts are found in the Valley of the Delaware under ten inches of such vegetable soil. Ten inches would represent the accumulations of about thirteen hundred years. The arrow-heads found at such depth are badly formed and not well pointed.

At the depth of two or three inches arrow-heads are found smaller in size, more symmetrical in form and neater in finish. They are between two hundred and fifty and four hundred years old.

The freshets of the Delaware, it is estimated, deposit 1-256th of an inch every year over the bottoms they overflow. Now bearing quartz, and spread out the freight of rubbish over the ravines and valleys below, entombing often the remains of the elephant and the mastodon. This glacial drift is the miner's "pay-dirt." It antedates the volcanic eruptions whose lava forms the Table Mountains and the high peaks of the Sierra. It antedates the great cañons which rivers have plowed into the hardened lava. Now the fragment of a human cranium has been found in the "pay-dirt" a hundred and eighty feet below the surface of the Table Mountain! A skull, covered with cemented gravel, has been found under five successive lava beds. In the upper lava bed there was not a crack through which a pebble could have fallen, and the gravel bed which held the skull held also the remains of a Mastodon. Man was in California when the site of Table Mountain was the bed of a river, and before the peaks of the Sierra had risen from the throat of a volcano. We have found him in a past so remote that geologic changes are its

hearths and shell heaps are found in the loess two feet below the present meadow-surface. Two feet of loess would represent the accumulations of 6,144 years. The implements found at the depth of two feet are crude and unpolished. Those found nearer the surface are of more symmetrical form and neater finish.

The same law holds as to pottery. Coarse, unornamented pottery is found associated with coarse, unskillfully chipped weapons. Finer pottery, made of carefully selected clay mixed with pulverized shells, is found associated with more polished weapons.

The inference is that the Indian appeared on the Atlantic sea-border about six thousand years ago; that he appeared first as a crude maker of arrow-heads and pottery, and that he improved slowly in tool-craft, and reached the limit of his powers about two hundred years ago. But the modern Indian came as a usurper. Implements dropped by men of the "rough Stone Age" are found in river gravel older than the loess.

only record. We wish to penetrate that past more deeply.

"The days of the years" of man's pilgrimage on earth have been divided into:

1st. The Rough Stone Age, or Palæolithic.

2nd. The Smooth Stone Age, or Neolithic.

3rd. The Bronze Age.

4th. The Iron Age.

This classification is based on *tools*, and it takes man where he first emerges from an unrecorded past and appears in the geologic record as a "tool using animal." He appears first as a maker of crude, ill-formed, unpolished implements of stone. He is lost to us for a while, and then re-appears as a more skillful workman, though still a workman only in stone. In later times we find him a worker in bronze. The Bronze Age closed just before history began. The first author who wrote in Europe—the first whose words have come down to us—speaks of the *ancients* as having used bronze, and not iron. One of the first writers of Asia speaks of a certain Tubal-cain as a *whetter* of bronze and iron; and further on, in his book of Deuteronomy, he speaks of a land whose stones are iron, "and out of whose hills thou mayest dig *bronze*."*

THE IRON AGE IS THE AGE THAT NOW IS.

No race is living now in a bronze age, but many savages are still in an age of stone. While we indicate the general progress of humanity by this classification,

* In these passages the English version reads *brass*. This is a blunder of the translators. In the second passage the sacred penman himself blundered. There are no hills where thou mayest dig either bronze or brass.

we are not to understand that Asia, Europe, Central America and Egypt passed from the use of stone to the use of bronze and left off bronze for iron at the same time. We say nothing of North America now, because she never got out of the Stone Age. The Mound-builder worked copper, it is true, but he worked it as a *stone*, not as a metal. Central America, at the time of discovery was in an age of bronze.

The peat-bogs, buried-lake villages, and caves of Europe record her Bronze Age. The older shell-mounds and tumuli hold a record of the Neolithic, or polished Stone Age. Of the transition from this to the Bronze Age we have no memorials, although it would seem the transition could not have been abrupt. Men must have used copper and tin separately before they learned to combine them and make *bronze.* The Mound-builder, indeed, was in a sort of transition between an age of stone and an age of bronze. He had copper but not tin.

The memorials of the Palæolithic Age we shall find in gravel-beds and caves.

Fig. 47.

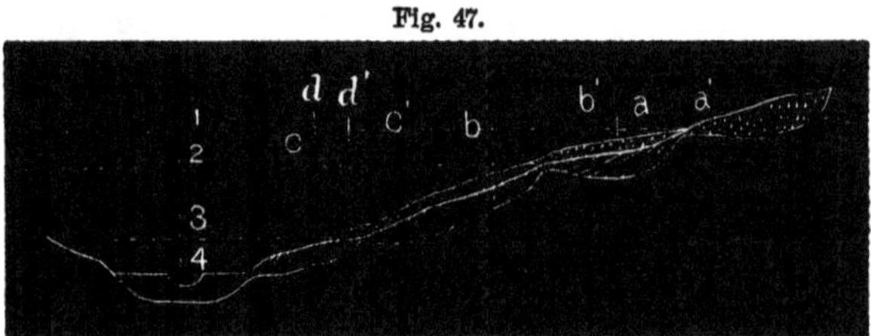


A river writes in its banks its own history and snatches of the history of animals and men who have lived in its valley. We cannot better show *how* a

Fig. 47. Ideal section of a river-valley.

river writes this history than by taking a section from Sir John Lubbock. The present position of the river is shown at 4 in Fig. 47. The earliest water-level is shown by the line 1, and the bank of the river while it flowed at that level is shown in the gravel-bed marked *a*. The gravel was spread out by the river at its usual flow, while the loam represented by *a* as overlying the gravel is the finer sediment dropped by the river in times of flood. A river eats its channel downward through the rocks, and the line 2 represents a later water-level. The gravel-bed marked *b* represents the river-bank while 2 was the water-level at the average flow, and the loam *b′* is the finer deposit at flood-time. Figure 3 represents a level after another period of erosion, *c* and *c′* and the gravel and loam as before. Finally, figure 4 represents the river at its present level, and *d* its present gravel-bank, and *d′* the loam of its present floods. The distance between 1 and 4, measured on the line 1, 4, is the depth of the erosion, the total work done by the river. In the rivers of France and England the length of this line is generally between fifty and a hundred feet. Erosion is a slow process, and the line, 1, 4, if a hundred feet long, must represent a very long time.

We have seen that man, before civilization had given him dominion over nature, clung to the shores of oceans and lakes and the banks of rivers. Now if man lived by a river when its level was at 1 and *a* was its bank, we should not expect to find memorials of him in shell-heaps. Too much time has elapsed for that. If he lived there at such time, and had so far emerged from a state of nature as to be a maker of tools, we might find memorials of him in his works.

Now in the highest river-gravels of the Thames, the Waveney, the Ouse, and the Avon of England, and the Somme of France, we find rough, ill-shaped implements of stone, the memorials of Palæolithic man. They differ greatly in form and finish from the implements of later age found in shell-heaps and tumuli and peat. They differ in material, too, for while the Neolithic man wrought in flint and slate and obsidian and sienite, the Palæolithic man chipped his adze, or hatchet, or scraper, or arrow-head out of flint alone. We cannot think that man passed suddenly from a rough worker in flint into a skilled worker in flint and slate and granite, but no tools are found which mark the transition, and the period which lapsed between the Palæolithic and Neolithic ages is an absolute blank. We talk of missing links in the chain of species. Here is a missing link in the life-chain of our own species. Who was the *Meso-lithic* man? How long did he live? What memorials did he leave, and where are they? The geologic record does not answer, and we pass from the Neolithic man who shared the earth with animals only a few of which have become extinct or changed their geographical range, to Palæolithic man whose mammalian companions have almost all become extinct or, migrating into other regions, survive in modified species. Where should we find *their* remains?

The deep recesses of a primeval forest are no more inviting to animals than men. They are solitudes. Birds even do not penetrate them. Mammals and birds alike fix their abodes on the skirts of a forest, and delight in alluvial plains and river-banks. Now in the upper gravel-beds of the Somme, associated with the flint implements, we find the remains of Elephants, Rhinoceros, Horses, Oxen, Cave Tigers, and Cave Hyenas, all of extinct species. The Elephant was the great woolly-haired Mammoth, and the Rhinoceros was the large, hairy, two-horned species. Its bones are found in this gravel, cut and gashed by crude stone implements. In the high gravel-beds along the rivers of the south and southeast of England we find, associated with unpolished instruments of flint, remains of the same extinct species; and in London, right down under Waterloo Place, under St. James's Square, under Bethel Green, and under Charing Cross, we find the remains of the shaggy Mammoth, the hairy Rhinoceros, the Musk Ox, the Hippopotamus, and, in the same gravel, the tools of Palæolithic man. In the shadow of a mighty past sits the great metropolis of the world! If the Neolithic man reared his hut on the site of our parvenu cities of the West, the Palæolithic man chipped his weapon of flint on the site of London, and warred against the hairy Rhinoceros and the Hippopotamus wallowing in the Thames. What ages lie between London and the Palæolithic man, whose weapons are in the gravel under her street we can better estimate after we have seen the record left by Palæolithic men in their *homes.*

More perishable have been the abodes of men than their tools, or even their heaps of refuse. Mounds of dust are alike the palaces and the huts of Nineveh, of Babylon, of Tyre and of Tadmor.

As war against hunger drove men to the water's edge, so war of man against man, as it drove the Venetians into the sea, drove Bronze Age men in Switzerland into the lakes, and Neolithic men in

America well up into the clouds. On the crests of mountains in New Mexico stand the untenanted stone huts of a tribe of Moquis Indians. A single misstep from the doorway would send you toppling over a sheer precipice down nearly three thousand feet.

> ". . On the crest that braved the sky
> A ruined hamlet towered so high
> That oft the sleeping albatross
> Struck the wild ruin with her wing,
> And from her cloud-rocked slumbering
> Started — to find man's dwelling there
> In her own silent fields of air!"

Older than the mountain-top huts of America are the compound Beehive houses of Scotland — older and younger, for they date from the Stone Age of Europe, and were tenanted, on the shores of Loch Resort, down to the year 1823.

Older than the Beehive house was the "Weem." It was the house most common during the Bronze Age

Fig. 48.

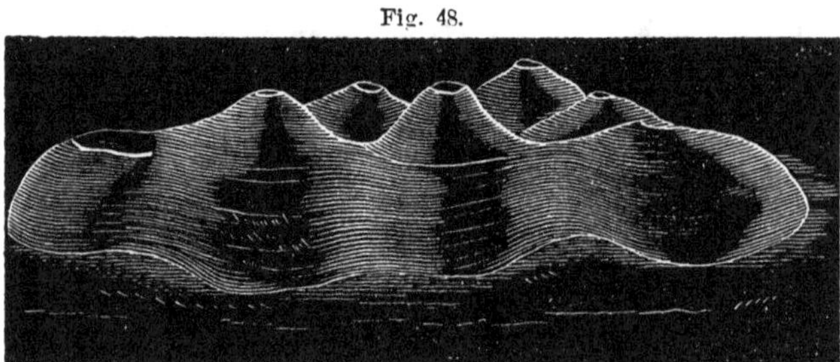

of Europe. It was subterranean, or semi-subterranean. Lubbock describes this house as simply a pit dug in the earth and covered over with boughs, and

Fig. 48. Bee-hive houses.

he tells us that on almost all large tracts of uncultivated land ancient Weem villages may be traced.

The Weem would seem to be a transition from the cave to the house, and it takes us back very far toward the times when man did not *make* his dwelling, but *chose* it.

In Archæan times, as we have seen, the atmosphere was loaded with carbonic acid, which made it an agent of destruction to all rocks containing alkali. In our times, as experiment has shown, the air is almost free from this acid, and the rocks suffer but little decay from their investing atmosphere. The foe which used to lurk in the air lurks now in percolating waters, and water *under* the surface works destruction as the air did *on* the surface. Rain water, sinking through the soil, absorbs carbonic acid from decaying vegetation. After passing through the soil and drift it sinks down through cracks and fissures in the rocks. If the strata are limestone, the acid eats slowly into its walls, and the water, seeking a lower level, and following the course of cracks or fissures, emerges somewhere as a spring or stream of *hard* water. It entered the rock *acidulated* water, it emerges from it *lime* water, and the lime it carries is a measure of the destruction wrought by its acid on the alkaline rock along its subterranean path. The chemical force of the acid is aided by the erosive powers of the water. And as the waters on the earth collect in rills and brooks and rivers and lakes, so "the waters under the earth" find the little openings made along fissures and rents and by acid and erosion plow their way and become Stygian rills which unite into rivers, or collect in chasms eroded in earlier times and form Tartarean lakes. The source of every well or spring is an underground stream.

Sometimes the walls and ceilings of these eroded chasms are beautifully sculptured. The dome-shaped ceiling of Wier's cave in Virginia is spangled with rosettes of snow-white alabaster. The spacious halls of Adlesberg cave in Germany are festooned with glittering spar. Gigantic stalactites, pendent from the ceiling, have met stalagmites equally gigantic rising from the floor, and formed immense pillars of alabaster.

So the cave is made, and sometimes beautified, and when the water has forsaken it, nature hands it over to animals and men. Primeval man did not *make* his house but *chose* it.

Through many ages men stuck to their subterranean abodes. History, at her birth, finds the Jew still in his cave. "And Lot went up out of Zoar; and he dwelt in a cave, he and his two daughters." Petra, "a city hewn in the rock," was a sort of "Weem village" which marked the transition from subterranean dwellings to subærial. The most ancient temples of Asia were built at the mouths of caves and in architectural imitation of caves. The Delphic oracles were uttered at a cave's mouth, and an English cave is known to this day as

"The blood-stained mansion of gigantic Thor!"

Religion, like royalty, dresses herself in archaic forms and manners, and, as the interdiction of metal at the altar by Egyptian and Jew, leads us to hypothecate a Stone Age back both of Pharaoh and Moses, so cave-temples and cave-oracles and encaved gods, would lead us to hypothecate, back of all this, simply cave *men.*

"*Specus erat pro domibus.*" *

After men had become civilized — or thought they had — dread of the supernatural deterred them from exploring the caves. At last one superstition drove out another. In the sixteenth and seventeenth centuries "Unicorn's horn" ranked high as a medicine. And where should the fabulous horn of the fabulous Unicorn be found? In the caves. In Germany and Hungary and Franconia, caves were explored for the potent drug. Teeth and tusks of Elephants, the teeth and bones of Lions, Tigers, Hyenas, Cave-bears, and even the remains of Men were all "Unicorn's horn" and supplied the materia medica of Europe with the most potent of its curatives.

From the account we have given of the origin of limestone caves it will be seen that they are connected with the drainage system of a country in which they occur. Before we can understand the relative age of its treasures, we must know whether the cave we are exploring stands in relations with the present drainage, or with some past system of drainage.

We give an ideal section through a chain of caves opening into a ravine which widens into a valley. We enter the first cave through a little pit and pass from gallery to gallery, from chamber to chamber, bewildered often by scenes of matchless beauty — floors studded with bosses of spar, red or snow-white, and ceilings festooned with stalactites which hang in deli-

* Pliny, His. Nat. 1, v.

cate tassels and even reach the floor in hollow shafts of alabaster no larger than straws * — we pass on from chamber to chamber till we emerge into the ravine at *r*. Passing down the ravine we find that it merges into gradually a valley. The caves are the channels of a subterranean stream which debouches into the ravine

Fig. 49.

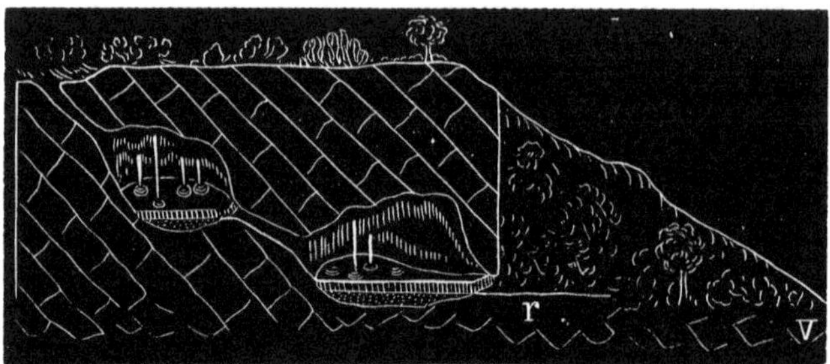


at *r*. Now as the valley is encroaching on the ravine, so the ravine is encroaching on the caves. For the cliff at the cave's mouth is weathering and crumbling into rubble which falls down and is swept away by the stream. *A ravine, then, is simply a cave unroofed.* And as the valley encroaches on the ravine, *a valley is simply a ravine whose walls Time has chipped off.* Cave, ravine, valley, this is the order in time, and we may infer that all the limestone valleys, the world over, began in subterranean caves, and that the caves, losing their roofs, became what a French geologist has called "*caves open to the sky,*" and that the ravines weathered finally into valleys.

The caves we have drawn in section would be con-

* The Cave of Caldy. See Boyd Dawkins on "Cave-Hunting."

Fig. 49. Ideal section of Caves, Ravine and Valley.

nected with the drainage and present geographical features of their district.

We give in Fig. 50 a tranverse section of a cave,

Fig. 50.

and ask the reader to imagine it placed in a series of caves like that shown in Fig. 49, but dry and old and brimmed with rubbish.

We will let Mr. Pengelly, from whom we have taken it, give us its story. We need not be told that it is the segment of an extinct water-course.

Fig. 50. A tranverse section of Brixham Cave.

"Tributary streams brought in the rounded pebbles, *C*, which fill the bottom of the fissure. It was only during occasional droughts that the cave seems to have been frequented by animals, their remains being scarce in the gravel bed, while the indications of man are comparatively numerous. . . . The cave became drier and was more resorted to by predatory animals who carried in their prey to devour, and was less frequented by man. . . . The cave would long continue to be subject to inundations the muddy waters of which deposited the cave earth, *B*, burying progressively the bones left from season to season by succeeding generations of beasts of prey. During this time the occasional visits of man are indicated by the rare occurrence of a flint implement lost, probably, as he groped his way through the dark passages of the cave." In this cave-earth, *B*, occur the bones of the great Mammoth, the Woolly Rhinoceros, the Lion, the Cave, Grizzly, and Brown Bears, and the Reindeer. Three of these species are extinct, and the Reindeer has left England for the North and the Lion for the South. We will give our attention again to Mr. Pengelly? "As with the change of climate the floods became less, so did the cave become drier, and was more resorted to by animals. At last it became a place of permanent resort for Bears. Their bones in all stages of growth, even sucking cubs, were met with in the upper part of the cave-earth (*B*) in greater numbers than were the bones of any other animals.

"Finally, as the cave became out of reach of the flood waters, the drippings from the roof which, up to this period had been lost in the accumulating cave-earth or deposited in thin incrustations on the exposed bones, now commenced that deposit of stalagmite, *A*, which sealed up and preserved undisturbed the shingle and cave-earth accumulated under former and different conditions. The cave, however, still continued to be the occasional resort of beasts of prey; for sparse remains of the Reindeer, together with those of the Bear and Rhinoceros, were found in the stalagmite floor. After a time the falling in of the roof at places, and the external surface weathering, stopped up some parts of the cave and closed its entrances with an accumulation of débris. From that time it ceased to be accessible except to smaller rodents and burrowing animals and so remained unused and untrodden until its recent discovery and exploration."

Near Stanhope, in England, at the end of a ravine, is a cave called "Heathery Burn." It is partially traversed by water which trickles from the mouth into a little rill called Stanhope Burn. As the cave is still the subterranean channel of a living stream that flows through a ravine into a valley, it cannot be geologically old. A stratum of sand on the bottom of this cave, sediment deposited by the stream, is covered by a sheet of stalagmite three feet thick. This "drip stone" is the accumulation of a period during which the stream must have been flowing through another channel. On the floor of stalagmite, bones and implements were found imbedded in sand and covered by another floor of stalagmite. A human skull has been found between this floor and roof of stalagmite, and bone pins, and *bronze* pins, and knives and armlets and spear-heads, and the broken bones of the short-horned Ox, and Badger and Dog, and bits of charcoal. The charcoal and broken bones show that the cave had been used as a habitation by men of the Bronze Age.

Bronze implements have been found in a few other caves of England, as in Thor's Cave, Victoria Cave, and in the caves of Cartwell, and in some caves of the continent, but they are not common. The cave had so far ceased to be important during the Age of Bronze that it throws but little light on times so near our own.

Near Corwen, in Wales, is a number of caves whose mouths open close by a shell heap. The implements found in these caves are all of polished stone, and the bones are those of animals which are still living and on the same parallels of latitude.

Broken bones of bears and wolves and dogs showed that these animals were eaten by Neolithic men. We have abundant evidence from other caves, and from tumuli and shell-heaps, that dogs were used as food during the Neolithic Age. Schweinfurth tells us that in Africa, wherever he found a tribe eating dogs, he found that such a tribe had recently left off cannibalism. This is their way of "letting down." We find Neolithic man always in the *dog* stage, never, except through stress of famine, in the *cannibal* stage.

Caves inhabited by Neolithic man are found all over Europe, and wherever they have been explored they have yielded substantially the same testimony. A small, swarthy, black-eyed, oval-faced people, making implements of stone and polishing them, with dogs, and swine, and little short-horned cows, which they had tamed and brought from Asia, had overspread Europe and was dwelling in her caves.

Near Torquay is "Kent's Cavern," whose fame for fifty years has filled the world. The entrance to this cave is on the side of an isolated hill, and not as the entrance to a cave connected with a present water-course, at the end of a ravine. The cave consists of galleries and chambers which have been explored with the greatest care by a number of English geologists under the leadership of Mr. Pengelly. We give, in Fig. 51, a section through one of these chambers. In the ascending order the deposits are, number 1, breccia and red loam; 2, crystalline stalagmite; 3, red cave-earth; 4, stalagmite; 5, black mould; 6, blocks of limestone cemented together by stalagmite.

Fig. 51.

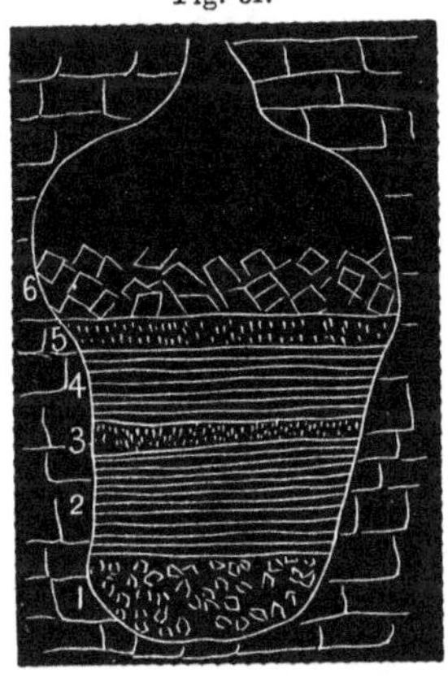


The blocks of limestone of course have fallen from the roof. We explore the black mould, marked 5, and find it to contain human bones and fragments of pottery, and implements made of stone and bronze, and barbed spear-heads of iron. The bones which occur in this mould are those of the Deer, the Ox, the Sheep and the Pig. We cannot ascribe a great antiquity to these remains, as the animals were those which are still indigenous to the country, and the bronze and iron implements could not have antedated the period of the Roman possession.

Under this mould, which is the record of recent

Fig. 51. Section of Kent's Cave.

times, we come to a floor of stalagmite No. 4 of our section. It varies in thickness from a few inches to five feet. Names cut in this stalagmite two hundred years ago at a spot where the drip is most copious and the thickness is greatest, are still visible. The drip has deposited, in two hundred years, a leaf of stalagmite not more than an eighth of an inch thick.* This rate would give a foot in twenty thousand years. Now when we remember that since men dropped their bronze instruments in the cave, say fifteen hundred years ago, no continuous sheet, but only patches of stalagmite, have been formed, and when we couple this fact with the rate given by two hundred years of drip over a single spot, we can form some conception of the time required for the formation of a continuous floor of stalagmite from three to five feet thick, and holding this conception firmly before the imagination, we shall not be disappointed at what we find *below* the stalagmite. The bed of cave-earth (No. 3) underlying the stalagmite is sediment thrown down by floods when these galleries were the channel of a subterranean water-course. But where did the stream begin? Into what ravine did it debouch? The caves are in an isolated hill and their mouths are seventy-five feet above the surrounding plain. When these caves were the underground segment of a brook, the surrounding country must have stood at least as high as their level, and the hill was *not* a hill. The geographical features have changed, and with them the drainage. What of climate and life? In this Cave-earth we find many bones, but not the bone of a dog, a pig, a sheep, a goat, or an ox. The species which left their remains

* British Association Report, 1869.

in the black mould above the stalagmite, and which characterize England now, are conspicuous by their absence. But here are the remains of the Cave Hyena, the Cave Bear, the Reindeer, the Woolly Rhinoceros, and the great Shaggy Mammoth. These animals are extinct, all except the Reindeer, and this is extinct in England. The Rhinoceros with its coat of hair, and the Mammoth with its reddish coat of wool and hair weighing three thousand pounds, were fitted

Fig. 52.

for the arctic cold which delights the Reindeer. In this Cave-earth we are looking on the memorials of a very different England from that of to-day. England has passed through a triple change of climate and fauna and physical geography since Kent's Cave was a den of Hyenas and Cave Bears, which dragged into their lair the dead Reindeer and Rhinoceros and Mam-

Fig. 52. Mammoth.

moth. The hundred thousand years or more recorded in the floor of stalagmite are a time-scale none too long for the measurement of changes so vast. But *man* was already in England—a troglodyte, housed with hyenas and bears. His crude implements of flint are strewn in profusion through the same cave-earth that holds the remains of the extinct fauna.

Below the cave-earth is another floor of stalagmite, (No. 2) indicating a period during which no water was flowing through the cave. This lower stalagmite is crystalline and, in places, twelve feet thick. More time would be required for the formation of this floor than the upper one. It has yielded a few bones of the Cave Bear. Under it we come to (No. 1) a deposit of breccia and red loam. In this we find the remains of the Cave Bear, an extinct species of Lion, a Hippopotamus, and the unpolished flint implements of Palæolithic man!

On the day Victoria was made queen an event was transpiring in Yorkshire of greater magnitude to the world of thought than the pageantry of coronation in London. A Mr. Jackson was crawling into a little opening he had discovered on the face of the cliff near Settle. That cave is likely to become the most famous in all the world, but it will bear the name of Victoria instead of Jackson.

Victoria cave consists of three large chambers, filled with debris almost to the roof. At the entrance is a large overhanging cliff, indicated by *d* in our section (Fig. 53). Time has chipped off fragments from the cliff, which have fallen down and formed a talus (No. 1). Through this talus we go down two feet and find a black layer (No. 2) composed of fragments of bone and charcoal and burnt stones—hearthstones they had been—and iron spear-heads and nails and daggers and finger-rings and armlets and bracelets and beads of amber and brooches of bronze and the ivory boss of a Roman sword and silver coins of Trajan. Some time about the middle of the fourth century a people who had learned civilization from Rome, under

Fig. 53.

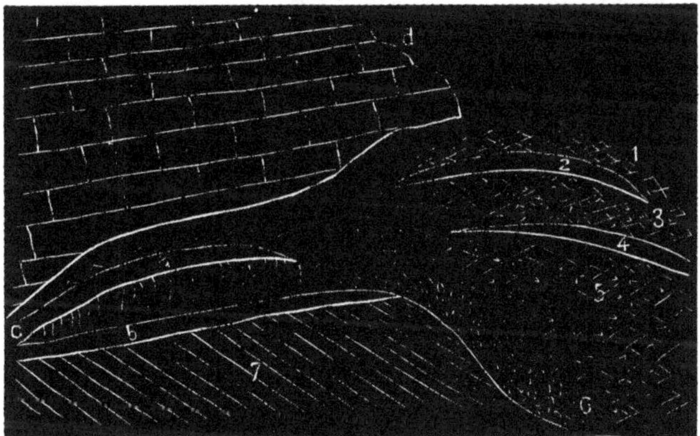


Fig. 53. Section of Victoria Cave.

the pressure of some great calamity, had sought refuge in this cave, and hither had brought their goods and their moneys. The overlying talus is two feet thick, and represents the disintegration of the cliff through a period of about fifteen hundred years. The underlying talus, 3, is six feet thick, and if the climate was the same during its formation as it has been for fifteen hundred years, it would represent the chippings of four thousand five hundred years. Under this talus, in the bed marked 4, we find the remains of Neolithic men. Below this the scientists who for six years have been exploring the cave have passed through nineteen feet of talus (No. 5) in which they have found no traces of man. But they have found at the base of this talus a number of glacial-worn bowlders and a bed of stiff glacial clay (No. 6), containing scratched bowlders. This clay enters the cave and is spread out over a portion of its floor. It is the inscribed record of the Age of Ice.

The cave has not yet been thoroughly explored, and we indicate in the section only two horizons, *a* and *b*. The upper bed of cave-earth (*a*) lies close to the roof of the cave and contains teeth and bones of the Grizzly Bear, the Red Deer, and the Reindeer. The presence of the Reindeer would indicate a cold climate.

The lower bed (*b*) extends from the glacial clay which it underlies, backward until it blends, as at *c*, with the upper bed. It holds the remains of the Hippopotamus, the Rhinoceros, (not the hairy,) the Elephant, (not the woolly,) the Hyena, and *Inter-glacial man!* The fragment of a rough, clumsy human fibula has been found in the lower bed where it was overlaid by the stiff glacial clay. This bed is either pre-glacial or it is the register of a warm period intercalated between periods of glacial cold. The climate was warm enough for the Hippopotamus. It cooled down, and when the upper bed was forming it was cold enough for the Reindeer.

Older record of man than this has not been found, unless it be in Table Mountain and the gold drift of California. Older record we can hardly expect to find in caves, for caves themselves are perishable and nature has set a limit to their age. Every ravine whose walls are limestone is the work which time has done in unroofing and destroying caves. Every valley in limestone is the work which time has done in chipping the walls of a ravine into slopes. Where a ravine is, there, in times not very remote, caves have been. Where a valley is, there, in the abysses of the past, caves have been. Look where we will, the geologic record is in fragments. Kent's Cave is only the remnant of a chain of caves whose containing rock has been eroded and carried into the sea. The Thames is known to carry into the sea about eight hundred tuns of lime every day, or fourteen thousand tuns from every square mile of its water-shed every hundred years. Mr. Prestwich has estimated, at this rate of erosion, that a stratum one foot thick would be stripped from the Thames water-shed and carried into the sea in the astonishingly short period of thirteen thousand two hundred years! How long before the eroding waters shall have stripped off Brixham and Kent and Victoria, as they have peeled away so many of their brethren?

"Out upon Time!
He will leave no more
Of the things that are
Than the things that were."

He has left enough. We lift the curtain he has woven and look far back on our fathers and fatherland. England is a land of Elephants, of gigantic Tigers, of River-horses, of Grizzly Bears, and Hyenas. The climate is warm winter and summer, for the Hippopotamus must wallow in rivers not clogged with ice. Hyenas are prowling about their dens and trooping after a disabled Lion or Elephant or Rhinoceros until its strength has ebbed, when the cowardly assassins fall on it and drag it to their lair. They devour its flesh and crunch its bones. They are cannibals and devour each other, as the gnawed Hyena-jaw of the

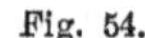

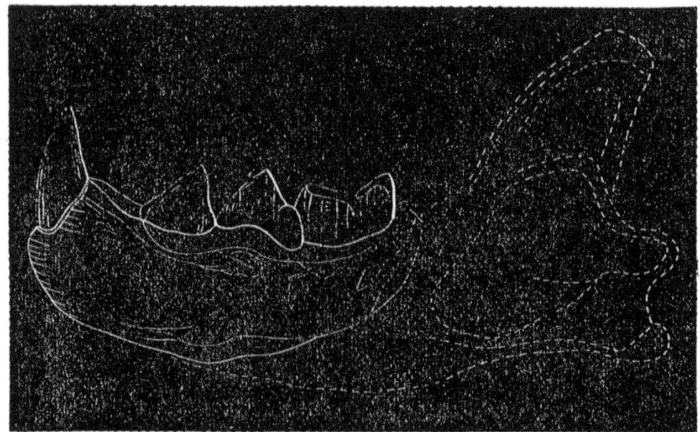

plate will show. There appears on the scene, from time to time, a being erect but miserable. Armed with bow and spear, man sallies out to attack his enemies. He drives the Hyena from her den and makes it his own. At the cave's mouth he kindles a fire, which serves the double purpose of cooking his food and keeping away his foes.

Time passes and the scene has changed. It is no longer England with the climate of Africa. The Rhinoceros is here, but it is clad in a vesture of hair. The Elephant is here, but it is dressed in a winter robe of wool. The Reindeer is here where the Hippopotamus had been. It is England in the climate of Siberia. Hyenas have survived the change and are still here; still prowling about the same den; still trooping after the sick and disabled; still dragging the dead and the dying into their lairs and devouring

Fig. 54. Gnawed under-jaw of a Hyena.

their flesh and gnawing their bones, as the gnawed Rhinoceros showed in the plate will attest. And man too had survived the change and is still here: still the same miserable savage, armed with the same bow,

Fig. 55.

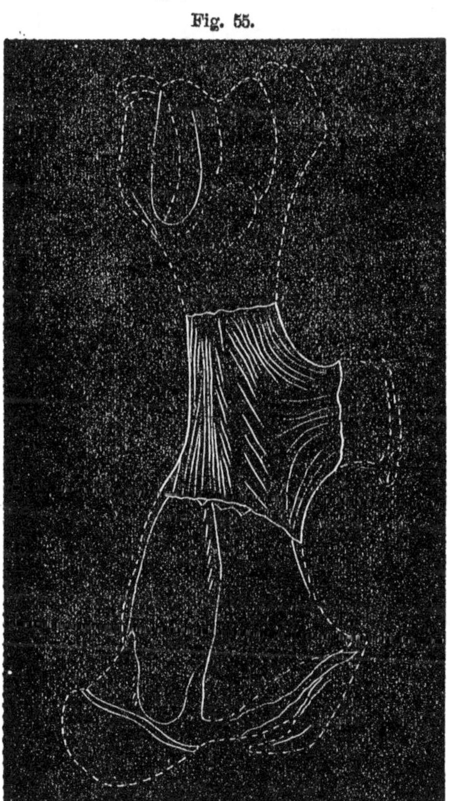

and the same lance and knife of flint. But he is no longer naked. The bone needles found in his caves show that he clad himself in robes of skin. His

Fig. 55. Gnawed Thigh-bone of a Rhinoceros.

country is large. From the site of London, where he drops his flint in the Thames gravel, he can walk on chalk downs over the Solent to the Isle of Wight, and over the English Channel into France. He has seen this land so warm that the Hippopotamus could bathe in its winter streams; he sees it now so cold that its high hills in summer are capped with snow and its ravines are gorged with glacial ice. A doom awaits him whose approach he does not see. Slowly, very slowly, over vast areas the face of the earth goes down, and Northern Europe is drowned under an ocean whose shells are left two thousand three hundred feet high on the Welsh Mountains, and whose floating icebergs drop their bowlders on the roofs of man's already ancient abodes.

Slowly the drowned land emerges, but the curtain does not lift again over Palæolithic man. *He* is gone, and science now echoes the despair of the historian, "lost *is* lost; gone is *forever* gone." The drowned land emerges, and over it spreads a race of men more skillful in tool-craft. The germ of civilization was in the Neolithic man, for he had domesticated the dog and the ox. The germ of religion was in him, for in his dead he conceived of something *not* dead, and he buried weapons to be borne away to an unseen land where—

"By midnight moons, o'er moistening dews,
In vestments of the chase arrayed,
The hunter still the deer pursues,
The hunter and the deer a shade."

He bequeathed to his children, spread over every land on the globe, a legend of the deluge of waters which had engulfed the land of his fathers, and the legend of his overspanning iris of hope. We have groped through his caves, and our torch of hope has burned brighter in the gloom. For from such a past has come the present, and from the present may unfold what a golden future! I have groped through his cave-homes and—

"The centuries behind me like a fruitful land reposed;
And I clung to all the present for the promise that it closed;
And I dipt into the future far as human eye could see,
Saw the vision of the world and all the wonder that would be."

# CHAPTER VII.

Origin of Animals — Magnitude of the Problem — Embryotic History of a Bird and Reptile — The Mud-fish, a Link between the Gill-breather and the Lung-breather — Transition from Mesozoic to Neozoic Times —Remoteness of the Eocene Age — Eocene Mammals — History of Teeth, and Origin of the Insectivorous Orders, of Herbivorous Orders, of Carnivorous Orders — History of the Feet and Origin of the Plantigrades, of the Digitigrades, of the Solipeds — Ontogeny and Phylogeny of the Deer — Creation by Hunger and by Love — The World is Poor — In Creation by Hunger, two Factors, Variation and Inheritance — Instability in Inorganic Nature implies Instability in Organic Nature — Advance and Recession — Degraded Types — Retrogression through Parasitism — Checks, Counterpoises, Stimulants —How They were Adjusted in Asia, in Australia, in South America — The Results.

THEY say that in the portals of the temple of knowledge there sits a veiled Sphinx who puts a riddle to every one who would enter. And they say that this riddle no man can answer, although every man must attempt.

To-day the sphinx is sitting at the portals of science and putting her riddle to every votary — this wondrous world of life, how did it come to be?

Two thousand years ago a Hindoo Seer gave answer in a couplet.

> " Creative thought and passion in a cup
> The meditating Brahm once hurled,
> And when the seething foam had all dried up,
> The sediment was this bright world."

More than two thousand years ago a Hebrew Seer gave answer in a hymn of creation which has passed into the literature of every book-making people on the earth.

"In the beginning Jehovah Elohim created the heavens and the earth."

These words, whether of the Hindoo or the Hebrew are no answer to the Sphinx. *How* were things created? *How* did plants, animals, and man come to be? It is our boon to grapple with a problem, the greatest that ever engaged the thoughts of men. It is a problem with which the religious sentiment has nothing to do and for whose solution the only requirement is a knowledge of facts and mental integrity in the use of them.

The Dinosaurs of the Oolite and Cretaceous and the Birds of the later Cretaceous have given us the pedigree of the Bird. Its immediate ancestor was a Bird-lizard, its remote ancestor a Lizard. If any evidence were wanted as to the history of the class it may be found in the history of the individual.

A Bird begins as every other vertebrate and there is a time when you cannot tell whether the forthcoming animal shall be a Reptile, a Bird, or a Mammal. But the embryo Bird, after evolution has carried it away from the Mammal, still bears a close resemblance to the Reptile. We place on the page in Fig. 56, a representation, after Haeckel, of an embryo Chick and an embryo Turtle — the Chick in the eighth day, and the Turtle in the sixth week. You can hardly tell which is the forthcoming Bird and which the forthcoming Turtle. The limbs are almost alike. What is to be a wing is still disguised under the form of a leg. In eye, mouth, ear, nose, beak, brain, head, the one is almost an exact copy of the other. A little divergence is seen in the length of the tails and in the

Fig. 56.

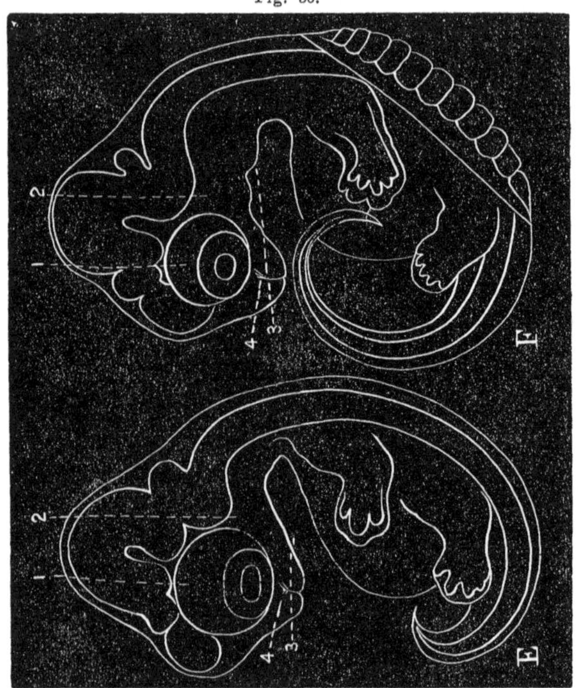


outlining of a tortoise-shell. If we had the embryo, not of a shell-bearing but a scale-bearing Reptile, we would see that scale and feather begin alike in little bosses raised on the skin.

In Fig. 57, we give two other embryos, one of a

Fig. 56. EMBRYOS. E, of a Chick; F, of a Turtle; the Chick in the eighth day and the Turtle in the sixth week. 1. The Eye; 2. The Ear; 3. Gill-arches; 4. Mouth.

Turtle four weeks old, and the other of a Chick four days old. The segmentation of the spinal column is the same in each. The limbs are the same. They appear as little fleshy knobs pushed out from the

Fig. 57.

body. In this early stage of development a very remarkable structure appears in each embryo. It is *gill-arches.* If these were fish-embryos, nature's meaning here would be obvious at a glance. But they

Fig. 57. Embryos of a Chick and Turtle, the Chick in the fourth day and the Turtle in the fourth week. Figures the same as in Fig. 56.

are embryos, the one of an animal which is to walk over the ground, and the other of an animal which is fly through the air, and each to breathe by lungs, not gills. Why then these gill-arches? Where is the mind which creates the meaningless? What architect builds only to raze his half-built structure to the ground? But these gill-arches are *not* meaningless. They are heirlooms from the Fish. As the Bird came from the Reptile, so the Reptile came from the Fish.

Fig. 58.

In the warm rivers of Africa and South America we find a peculiar race of vertebrates called "Mud-fishes." The body is covered with overlapping scales and provided with long ribbon-like fins pointed at the extremeties. The Mud-fish has lungs which extend through the greater part of the body and two sets of gills, one external and the other internal. In the adult the external gills are atrophied. The Mud-fish breathes alternately water and air and presents nature

Fig. 58. The Mud-fish.

in transition from a gill-breather to a lung-breather. The Mudfish is a *fish* by right of spinal column and limbs. By right of skull and lung and heart it is a Reptile.

The transition from Fish to Reptile was accomplished during the palæozoic age; that from Reptile to Bird toward the close of the mesozoic age. From the mesozoic we pass to the neozoic and find ourselves at once in the age of the reign of Mammals. We pass through what seems a great gap. The Permian system registers the transition from palæozoic to mesozoic times. On what strata was recorded the transition from mesozoic to neozoic times?

The estuary deposits of the Wealden may be traced from a point a few miles northwest of London to Beauvais in France. We have seen what gigantic Dinosaurs prowled along the shores of that estuary and the great river which brought to it sediment from a land we cannot locate. After the Wealden age the land went down and all traces of land-life are lost. When it emerged again, capped with chalk, the site of London was another estuary. On its bed was spread out the strata of London clay, five hundred feet thick. Looking back into the mediæval age of earth-history, we see the land of the Dinosaur going down, slowly down, until it is drowned under a deep ocean. After the lapse of the Cretaceous period we see it emerging, but it emerges a *new* world — the *neozoic world*. The Wealden estuary and the London estuary, so near together are yet so far apart! The land whose rivers flowed into the old was the abode of Dinosaurs. The land whose rivers flowed into the new was peopled by Mammals akin to the Opossum, the Tapir, and the Hog. Whither went the Dinosaurs? Whence came the Marsupials and Pachyderms? Where was the unrecorded land of the ages between the two estuaries? What was its life?

Everywhere, between the close of the mesozoic and opening of the neozoic we find in the geologic calendar breaks like this — everywhere except in the Rocky Mountain region of North America. Hayden writes in the Geological Survey of Colorado, "there is no proof, so far as I have observed, in all the western country, of true non-conformity between the Cretaceous and lower Tertiary beds, and no evidence of any change in sediments, or any catastrophe sufficient to account for the sudden and apparently complete destruction of organic life at the close of the Cretaceous period." The highest authorities on fossil botany have pronounced the extinct flora entombed in a group of these western strata to be *Tertiary*, while all geologists would pronounce the extinct fauna to be *Cretaceous.* The facts show an uninterrupted succession of plant-life across what in other parts of the world has been a great gap between the mesozoic and neozoic ages. Plants crossed the gap, and some of the humbler species of Reptiles, but not a Dinosaur, not a Pythonomorph, not a Pterodactyle, not one of the complex Saurians characteristic of the mesozoic age. During that age only a few mammals had appeared and these, humble in structure and small in size. They were marsupials and insectivora. On the emergence of land at the close of the Cretaceous period and Mesozoic age, Mammals appeared as the dominant class. It is now the age of the reign of Mammals.

The oldest of the Tertiary formations is called

"Eocene," dawn of the recent orders of life and the recent order of things. But our minds must not be abused by words. When we say Eocene we must not *think* Eocene, unless our thoughts run along the geologic time-scale. For recent is not recent in any calendar save the geologic. Eocene rocks, formed on the bed of a sea, cap the Coast Range Mountains of California, where they attain a thickness of three or four thousand feet. They lie on the summit of the Dent de Midi, the Apennines, the Carpathians, the Caucasus, and the slopes of the Himalaya to a height of sixteen thousand five hundred feet. Since "the dawn of the recent," the Coast Range of California, part of the Alps, and all the Apennines have risen from the sea. The Himalaya have risen more than sixteen thousand feet. Recent? We do not think of the pyramids as recent, but during the Eocene period the very rock of which the pyramids are made was not itself made. The "Nummulitic limestone," built so largely into the pyramids, was in process of making The myriads of "Ray Streamers" floating on the bosom of an Eocene sea, were abstracting lime from the water and encasing their structureless bodies in coin-shaped shells which, at death, were falling to the bottom and forming Nummulite limestone.

The Mammals of the Eocene are of commanding interest.

If we had a museum illustrating the history of stoves, we would find that, in general, the further back we go in the series the nearer we approach to the old-style "fire-place." The old Franklin stove was simply an iron fire-place set out from the wall. Each successive invention or improvement was a departure further and further from this primitive pattern. So the creative mind of man has wrought, and *must* work. The latest stove or the latest steam engine would have been an absolute impossibility at the start. It implies a line of ancestral inventions. The old Franklin might be called a "generalized" stove, and the modern base-burner a highly "specialized" stove. We are attempting to set forth very complicated things in nature by things comparatively simple in art.

The simplest dental system a Mammal could have would be teeth all alike and evenly planted around the jaw. Such dentition we nowhere find. Differences in the kind of teeth occur in all terrestrial Mammals, living and extinct. Even the Eocene Mammal presents three kinds of teeth, *incisor*, *canine*, and *molar*. Molars, generally, are *grinding* teeth, canine *tearing*, and incisors *cutting*. The dental formula for most of the Eocene Mammals would be i. $\frac{3}{3}.\frac{3}{3}$, c. $\frac{1}{1}.\frac{1}{1}$, p. m. $\frac{4}{4}.\frac{4}{4}$, m. $\frac{3}{3}.\frac{3}{3}$, which is a convenient way of saying that there were six incisors in each jaw, two canines, eight premolars, and six molars — forty-four teeth in all. These teeth form an uninterrupted and even series. The crown of the molars of some of the oldest genera was quadra-tuberculate. This was the most generalized form of dentition among Mammals. Suppose that from an order of Mammals having such teeth there should in time develop an order of insectivorous Mammals. It would be of advantage to have the molar teeth raised into conical points, as they are in the Insectivora. Suppose there should arise an order of Carnivora. It would be of advantage to have the molars so compressed as to form cutting edges, and the canines long, curved, and pointed. We can trace the gradations from that generalized dentition to this specialized *carnivorous* dentition. The hindmost molars and the foremost premolars disappear, and finally, in the cat family, only one true molar is left in the upper jaw.

Suppose there should arise, in time, an order of ruminating Herbivora. Such animals seize the grass with their tongue, and crop it by pressing the lower incisors against the callous pad which occupies the place of the upper incisors. It would be of advantage to a Ruminant using its tongue for prehension to have the front teeth of the upper jaw suppressed. In the Ruminant they are suppressed, and in some Ruminants both incisors and canines are suppressed. The crown of a molar is not quadra-tuberculate. The tubercles had coalesced and formed ridges adapted for grinding grass. This is the *specialized* dentition of the Ruminant.

Suppose that a tribe of non-ruminating Herbivora were to rise, the animal seizing the grass with its *lips*, not with its tongue. Suppression of the incisors would be of disadvantage. The Horse, which is such a non-ruminant, has six long, curved incisors in the front part of each jaw. After the incisors comes a toothless interval, and then six large molars and premolars, modified in a peculiar way for triturating grass. This is the formula for the female. The male has a tusk-like canine adjacent to the last incisor. This dentition is more highly specialized than that of the insect-eater, the Tiger, or the Ox. We can trace the steps which led to it. We find the ancestor of the Horse in a *generalized* Mammal of the Eocene.

Suppose there should arise an order not specially

insect-eating, or flesh-eating, or grass-eating. Its dentition would not be specialized, but would answer more closely to the ancient, generalized formula. Such an order is that to which Man belongs Our teeth form an even and uninterrupted series. The crowns of the first and second molars of our upper jaw are quadra-tuberculate, according to the ancient pattern. The feet are in correlation with the teeth.

Taking the elements which enter into the composition of a foot, and drawing them in such form and position as we find them in the earliest of Eocene Mammals, we would have, in the lower segment of the limb, two bones, the tibia and fibula; in the highest segment of the foot two bones, the calcaneum and astragalus, or heel-bone and turning-joint. They are short and flat, and articulate, one with the tibia and the other with the fibula, forming a stiff and clumsy ankle-joint. Below these tarsal bones we have a row of five metatarsals, and attached to each of the five a row of toe-bones, three in each row. The toes terminate in little trowel-shaped hoofs. Such a foot would be *plantigrade.* The animal would have a slow and awkward gait, because it would have a stiff and clumsy ankle-joint. The elements of the foot would be shaped and arranged very much as in the Reptile. The foot would be a *generalized* one.

Suppose, now, that from an order of Mammals with such a clumsy ankle-joint and plantigrade foot, there should arise, in time, an order of fierce Carnivora. It would be of advantage to an animal, stealing on its prey, to have a light and airy tread. The heel-bone would be elongated and would cease to articulate with the fibula. The turning-joint would be modified to form a better articulation with the fibula. It would be of advantage to a flesh-eating animal to have sharp retractile claws, instead of harmless trowel-shaped hoofs. While the heel-bone and turning-joint are elongating, the metacarpals and phalanges elongate. The foot becomes *digitigrade* and the toes have hooked claws. Flesh-tearing claws are in correlation with flesh-tearing teeth. In the early Eocene most of the Mammals were *plantigrade*. In the following period, called *Miocene*, most of the Mammals were digitigrade.

Suppose there should arise an herbivorous order modified in such a way as to digest and assimulate prodigious masses of food. Such an order is the Ruminant. We have seen how its teeth were modified. The feet undergo a modification correlative to that of the teeth. Wherever we find the type of tooth peculiar to the Ruminant we may look for an astragalus, or turning joint, of the hind foot with two grooved and pulley-shaped surfaces, one above and one below, and a foot terminating in two toes, or if in four, the outer ones reduced in size. The reason for this correlation does not appear. We can see that with a grass-cutting tooth there should not be a flesh-tearing claw. If there is any modification of the primitive style of hoof it might be in the direction of *more* hoof. Our Ruminant would have hoof-bearing toes.

Suppose there should arise an herbivorous order modified for rapid motion over the ground. It would be of advantage to such an animal to have some of the elements of the foot suppressed and some greatly modified. The heel-bone would be elongated as in the Carnivora and the Ruminant. The turning-joint would present to the tibia the grooved segment of a pulley. The articulation would be so modified as to make a joint at once the strongest and most flexible. If the progress of the earlier ages was marked in the tails of fishes, one line of progress in the latter ages was marked in a certain foot-bone of Mammals. Exploring the Tertiary formations of the West, Cope found that he could determine the relative age of a group of strata by the fossil astragalus it contained. In the Ruminant one toe is suppressed and two are reduced,* and in this swift-footed order which is coming, Nature will move along the same path which led to the Ruminant, but she will move to her furthermost limit. In the Eocene rocks of Wyoming we find the ruins of a little animal which takes the name of *Oro-*

Fig. 59.

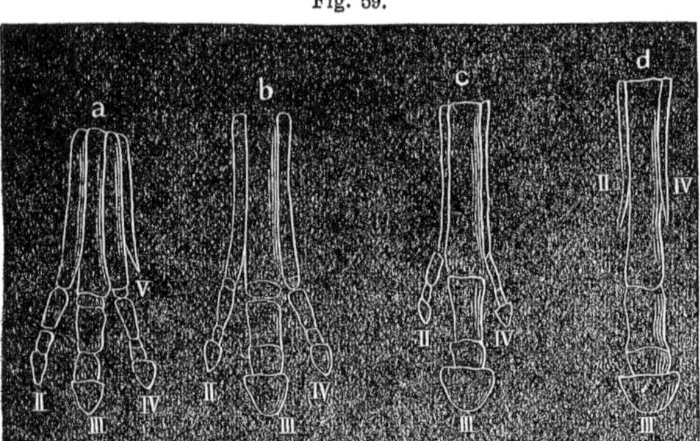


*hippus*, or Mountain-horse. It was hardly larger than a Fox. Our plate, Fig. 59, gives at *a*, the plan of its foot. It had three toes, representing the second, third and fourth of the normal five-toed foot. On the outer side of the foot appears a little style of bone which is

* In some species three are suppressed.

the vestige of the fifth toe. It is an heirloom from some ancestor which had five toes. Orohippus is a station on the road from a five-toed to a one-toed animal. In the Miocene rocks of Oregon and the Upper Missouri and the region of the Rocky Mountains, we find the remains of an animal closely related to the Orohippus of the Eocene. It is called *Anchitherium*. The plan of its foot will be seen at *b*. The style which in Orohippus represented the fifth toe has gone, every trace gone. The second and fourth toes are much reduced. They are evidently going the same way as the first and fifth have gone.

In the Pliocene rocks of Oregon and Niobrara we find the ruins of a later member of this series. *Hipparion* was closely related to Anchitherium. Its foot is represented at *c*. The middle toe appears now very much enlarged and the second and fourth toes correspondingly reduced. They were almost atrophied. They were of no possible use to the animal, as they did not reach the ground. Still they bore little hoofs like the "dew-claws" of the Reindeer.

Later down in time the modern Horse appears. Its foot is represented at *d*. The middle toe is enlarged still more than in Hipparion, and the second and fourth which have been going, going, going, are *gone!* A faint reminiscence of them is preserved in the "splint-bones." These bones are the remnants of the metacarpals, but their phalanges and hoofs have entirely faded out.

Orohippus could not have been the first of the series which led to the Horse, since its own foot had the vestige of a fifth toe, as a Horse's foot has the vestige of a second and fourth. The first member of the series must have been a five-toed, generalized plantigrade. The foot rested on the ground from heel to toe. The last member of the series is one of the most specialized of animal forms — in the foot the most specialized of all animals. All the toes are suppressed except the third, and nothing but its hoof reaches the ground. One toe has been selected and modified so as to form a great lever "which in combination with the levers constituted by the upper and middle divisions of the limb, forms a sort of double C-spring arrangement, and thus gives to the Horse its wonderful galloping power."*

We glance backward along the series and find that in its general features Hipparion was less specialized than the Horse. The ulna, although attached to the radius, was not anchylosed with it. The fibula, although anchylosed with the tibia, was not absorbed in it. Anchitherium was less specialized than Hipparion. As the ulna is more distinct in Hipparion than in the Horse, so it is more distinct in Anchitherium than in Hipparion. The same is true as to the fibula which is anchylosed with the tibia only at its lower extremity. The limbs of Anchitherium were shorter and its body larger and more clumsy in shape. Its dentition was more generalized. It had the forty-four teeth which belong to the typical mammal. The teeth were already patterned after the type they attain in the Horse, but their structure was simpler. The valleys between the ridges were not filled up with cementum as in the Horse. It is evident that as we reach backward along the series we

* Prof. Huxley.

are passing from more specialized to more generalized forms.

And now, as if to supplement all this evidence as to the derivation of the Horse, a Horse is found now and then with three toes, just as a child appears, but not so rarely, with some peculiar feature of a great great grandfather. It is *atavism*, or reversion to an ancestral type.

As we trace the idea involved in all stoves back through the minds of inventors to the old Franklin, so we trace the structures specialized in the Horse, the Ox, and the Tiger, back along convergent lines which unite in some generalized Mammal of a distant past.

Suppose, now, that from such a generalized Mammal there should spring an order not modified in tooth and toe for tearing flesh, not modified in tooth for cropping grass and in stomach for assimilating large stores of it, and not modified in foot and limb for rapid motion over the ground. To such an order Man himself belongs. His teeth keep very near to the primitive pattern. His foot is near the generalized primitive pattern. He is plantigrade. He has a short heel-bone, a short astragalus, and five short toes, terminating in what are neither claws nor hoofs, but nails half way between. His digestive system is near the primitive, undifferentiated pattern. Man is old-fashioned in foot and tooth and stomach. He is *new* and highly differentiated in hand and head.

It is obvious that specialized animals are more recent and the generalized more ancient. The Deer is more generalized than the Antelope and it came first in order of time. The Pig is more generalized than the Ox and appears earlier in the geologic calendar. The

Deer is separated widely from the Pig, but the Oreodon, a generalization of Pig and Deer, that is, an animal intermediate between the two, appeared earlier than either. The Elephant, in the structure of its limbs, is generalized, but in skull, teeth, and proboscis is highly specialized. The Mastodon was more generalized than the Elephant and it came earlier. The Dinotherium was more generalized than the Mastodon and appeared earlier on the geologic record.

A highly specialized Ruminant is the Camel. It differs from all other Mammals in having one strong incisor on each side of the upper jaw, and from all other living Mammals in having a peculiar structure of the neck-vertebræ which allows the artery that carries blood to the brain to pass up through the vertebral canal side by side with the nerve-cord. In the very young Camel there is a full set of upper incisors. In Miocene times an animal appeared in the region of the Rocky Mountains, so much of a Camel that it takes the name of *Procamelus*, or in English, *fore-camel*. It retained through life the milk dentition of the living Camel, and was, by so much, more generalized. Pro-camel might have become Camel by retardation in the growth of its teeth. In earlier times there roamed over the same region an animal not quite so camel-like, an animal called *Poebrotherium*. It was a little nearer the average Mammal than was its successor and descendant, the Pro-camelus. But already it had the specialized structure of neck-vertebræ which in after-time was to be a distinctive feature of the Camel.

The Lemurs are the lowest of Monkeys. In Eocene

times the Western slopes of America had Monkeys lower and more generalized than Lemurs.

Eocene times were characterized by the absence of Ruminants, Horses, and Elephants, by the presence of a few generalized Carnivora and Half-apes, and by the preponderance of generalized hoof-animals akin to the Tapir. Out of forty-five species of Eocene Mammals whose ruins have been taken from our Western strata and tabulated by Leidy and Marsh and Cope, twenty-eight were generalized.* Now science has found the radical of the Horse back among these generalized orders, and she has made out its pedigree. The demonstration does not admit the shadow of a doubt. She has indicated very clearly the origin of the Camels. She has traced the pedigree of Birds from animals which were generalizations of Bird and Reptile. She has demonstrated that Keeled and ribbed and tuberculated and spined Ammonites were developed from smooth and simple ones.

This is much. Suppose it were less. Suppose that the case rested on the Horse alone. What then?

When Newton had demonstrated the law of gravitation from atoms and apples and toppling towers, what did he do? He asserted *universal* gravitation. Men

* For a view of American life in Eocene times consult the frontispiece. The large, many-horned animal in the foreground is the Uintahtherium. Under Uintahtherium crouches an early member of the Cat family. To the right stands the Machærodus, one of the most specialized of Eocene Mammals. Further to the right and in the foreground is the Orohippus, and still further to the right the Pro-camelus. In the background is a primeval Pig. The progenitor of the Cranes is represented on the ground, and the primeval Eagles in the air. Pines, palms, and magnolias are the characteristic trees, and Lemurs, or Half-apes, the characteristic Simians.

came to him with objections and difficulties. Some were trivial and some were serious. The great man met them in a spirit something like this: "Gentlemen, some of your arguments are trivial. I would not answer them if I could. Some of them are serious. Just now I could not answer them if I would. Nevertheless, gravitation *is*, and it is universal." This was Newton's science. This is science herself, for she is based on conceptions of unity. If Newton had demonstrated the law of gravitation only from a falling apple, he would have been compelled by a law as deeply graven in the minds of men as gravitation in the face of nature, to assert universal gravitation.

So if the question of the origin of species rested only on what has been demonstrated as to the origin of the Horse, the biologist would have to say to all who question or doubt, "Gentlemen, you see the Horse. You see that it differs more widely from the average animal than man himself. Now we have shown that this animal came from an average animal, and was in process of creation through hundreds of thousands of years. When you press me with difficulties, and say that although my theory holds good as to the Horse, because I have not demonstrated it so clearly as to the Ox, therefore it fails, and all other animals have come by miraculous and instantaneous creation, I must say to you, 'The difficulties you urge may be real and in the present state of science I may not be able to meet them; nevertheless, evolution is true and its application universal.'"

Remembering now what the pedigree of the Horse has taught us, that evolution proceeds from the general to the special, the facts we have placed in array become luminous. Theory requires that the later species should be more specialized, and they are. But theory requires that if, as we press back into Eocene times, we near the beginning of mammalian life, all the species should be generalized, and they are not. The inference is that Mammals began long before the period we have called "The Dawn." The seventeen more specialized Eocene species must have had a mammalian ancestry reaching far back into the Mesozoic. In Mesozoic strata we find the remains of Marsupials and Insectivora. Where was the common stock from which these specialized classes diverged? In the Trias, the first member of the Mesozoic series, we find the remains of herbivorous and carnivorous Marsupials. From what common Marsupial stock did these species diverge? At the dawn of the Mesozoic age the lowest of Mammals must have been in existence long enough to have diverged from a common stock into two tribes, specialized, the one for eating grass and the other for eating flesh. The common mammalian radical must have been *palæozoic.*

The development of life on our globe, by whatever process, has proceeded very slowly from the general to the special, from the simple to the complex. It has not proceeded fortuitously, and species are not accidents. The history of the Horse is the exemplification of a law. The Ammonite from its stony sleep suggests a profounder law. Recall that rhythmic history — the life that rose in the individual as it rose in the race, that culminated in beautifying the individual as it culminated in beautifying the race, that declined and stripped the individual of ornaments as it declined and stripped the ornaments from its race, that made the individual a compression into a few years, of the race, and the race an expansion, through vast ages, of the individual. Very far from the Ammonite in time and in structure is the Deer, but if we turn from that dead sea-mollusk to this living land-mammal we find the same relations between the life of the individual and the life of its race. The earliest Deer known to us, a Deer of the lower Miocene, had no antlers. The next Deer, a Deer of the middle Miocene, had simple antlers with no more than two branches. The Deer of the upper Miocene had three branches to the antlers. The Deer of the next geologic age had a luxuriant growth of antlers adorned with many branches. So grew the *order* of the Deer. See now how grows the Deer. The young Deer has no antlers. It is a copy of its early Miocene ancestor. The Deer in the second year of its growth has antlers with two branches. It is a copy of its middle Miocene ancestor. The Deer in the third year of its growth may have three branches to its antlers and is then a copy of its upper Miocene ancestor. In maturity its antlers attain a luxuriance of growth like that which characterized the Deer of the Pliocene. As the order unfolded so unfolds the individual. A Deer lives over again in its own life, the life of a long line of ancestry. The law written in the shells of Ammonites and the antlers of Deer is universal though not everywhere patent. All things, Birds, Mammal, and even Man, repeat the past in the unseen and unconscious life of the embryo. The past is never forgotten. On all the lines of life it is carried up into the present, although it may be hidden from our eyes.

And now as animals have been created little by

little, and as the growth of the individual epitomizes the unfolding of the species, can science deduce from these facts an answer to the Sphinx and declare the *method* of creation?

> "Some day, Philosophy, no doubt,
> A better World will bring about,
> Till then the Old a little longer,
> Must blunder on through Love and Hunger."

The poet is always wiser than he knows, and in these lines Schiller has divined a great truth. The workings of the world are carried on by *hunger* and by *love.* Hunger prompts the struggle to sustain life, and love, the struggle to perpetuate life.

The world is poor. "Naked," said the patriarch of Uz, "naked came I into the world and naked must I go hence." Naked and hungry comes into the world every one of its children. Mouths are many; morsels are few. Everything is born to a heritage of hunger, or war, or both. The progeny of one Cod-fish, if they could find enough to eat and escape being themselves eaten, might within a few hundred years, brim the Atlantic from surface to abyss. Such prodigious fecundity and such prodigious sacrifice! Such hungry war in the sea, and so many lives devoured to sustain other lives! A single weed, unchecked, might cover the land within the time that a single fish was filling the sea. A single insect might, under the same conditions, within the same time, cloud the air and devour every green leaf as it opened to the summer. Such prodigious fecundity on the land, and such prodigious multitudes that starve or fall into hungry mouths and prevent others from starving! The flea that bites you is itself bitten. The "flea" that bites the flea is bitten by a biter still smaller. We have taken a mosquito gorged with our own blood and found her body covered with parasites which were sucking from her the blood she had drawn from us. Ourselves are not the first terms of the series nor the mosquito's mosquito the last. With the feelings we contemplated the punishment of the mosquito, with the same feelings the ox might have contemplated our punishment.

> "For Nature is one with rapine, a harm no preacher can heal;
> The May-fly is torn by the Swallow, the Sparrow speared by the Shrike,
> And the whole little world where I sit is a world of plunder and prey."

The world is poor. She proffers life to many and she sustains but few. She gives life but sustains it only as each living thing is up with bill, or beak, or tooth, or claw, or toilsome hand or sweating brow, to conquer the means of living. Organic nature is the war of each against all. "The whole creation groaneth and travaileth." We are warned against putting new wine into old bottles, but here is an old bottle that will hold all the new wine of Darwinism. "The whole creation groaneth"—and has ever groaned. No museum of lethal weapons made by the cruelty of man, could match those of the fossil world—"And travaileth." What is born of such travail? At the portals of birth is Pain. What if such travail were the birth-throes of species?

The successive modifications which led from a generalized plantigrade to the digitigrade Tiger affected chiefly the teeth and feet, fitting them more and more for fight. The successive modifications which led from the same kind of plantigrade to the Horse, affected the same parts, fitting the teeth better and better for cropping and triturating food, and the feet better and better for carrying the animal away from the cousins who would make it serve for *their* food.

In this process of creation by *hunger* two factors are involved, variation and inheritance. An organism is like a target against which nature shoots all her arrows. She assails it by hungry mouths of other organisms. She assails it by cold, by heat, by rain, by drought. Now an organism, if it continues to be, must keep itself very nearly in equilibrium with the nature which is about it. While nature without is changing from the cold of winter to the heat of summer, there must be responsive changes in the organism. While inorganic nature is unstable, stability in organic nature would be impossible, even if there was no mouth that hungered. Now whatever change occurs in an animal, that change nature is disposed to hold, and if of advantage, to put at usury. If she gets a sharper tooth or a stronger claw, or a swifter foot, she says, "go to, you are of use to me and I put you at interest." If among the Cephalopods that crawled over the bed of a Silurian sea with long conical shells on their backs, there appeared an individual with a shell slightly curved, such an individual would have the advantage over its fellows whose shells were straight. The curve would be likely to appear in the next generation. The individuals that inherited this twist would be better equipped in the struggle for life than the straight-shelled ones, and the individual whose shell twisted the most would be equipped the best. Thus the twist which appeared first as a sport became congenital, and after the lapse of many ages culminated in the involute coil of the Nautilus. And as in our own race the peoples who did not share the first departure on lines of progress became rigid and perished, so the early Cephalopods not affected by this movement, in the sharp conflict for life, were pushed to the wall and exterminated.*

Placing ourselves far back in the times of a radical, generalized Mammal, and taking our point of view from *within*, we can imagine such a Mammal, if endowed with intelligence and prevision, indulging in cogitations something like these:

"The times are changing and I must change. The face of nature through the coming æons will grow more complex. Time was when the world had no

* If the shell of the Orthoceras was badly shaped and Nature must either drop it or change it, how then did Nature get it? A very pertinent question which will rise in the mind of every thoughtful reader. Certainly not by "natural selection" as by this process Nature works for her own advantage. The Orthoceratites could not have arisen in times when they were pressed by enemies. They must have lived in comparative ease and safety. Their growth was a kind of vegetative growth consisting of mere repetition of parts. Their numerous arms, each a copy of its fellow, attest their low rank. The numerous septa of their shells are another badge of low, vegetative life. The shell grew on, adding septum to septum, each new septum being a repetition of the old. It grew to a prodigious length because its tenant was so safe and so sluggish. Even with such sluggishness and security the shell was inconveniently long, as we know by the fact that mature shells are always found broken off at the end. *Vegetative* life had gained the advantage and the animal became handicapped through immunity and laziness, its consequent.

features. The time is coming when the features that are will be intensified. Time was when far up toward either pole the world had a summer climate. The time is coming when differences between summer and winter will everywhere be intensified. Valleys will be deepened, mountains heightened, table-lands lifted up, regions of drought and region of moisture, regions of cold and regions of heat, will appear, and I, in the person of my descendants, must change with such changing aspects of nature. As the face of the earth becomes specialized into valleys and table-lands, and wooded plains and desert saharas, and zones frigid and temperate and torrid, my descendants, spread over the globe, must choose different habitats and be modified accordingly. Life must become more complex as its conditions become more complex. The struggle to sustain life must grow sharper as the range of any given type grows more restricted.

"Looking down along the changes occurring in the inorganic world I see a descendant of mine with teeth differing very slightly from my own. The change has come as a 'sport,' responsive in some way to the change occurring without. It is of advantage, for these children of mine are pressing each other. Some of them have taken very much to biting and over all other biters this fellow has the advantage. The sharpness of tooth will re-appear, with more emphasis, in his offspring. And those individuals in which it appears with greatest emphasis will have the best chance in the struggle for life. They will be most likely to make their way and perpetuate themselves in offspring. As carnivorous habits become more and more established, the teeth and feet will be modified more and more for carnivorous uses. I look far down along this diverging branch and see my posterity in the mottled Pard, in the jungled Tiger, and in that stealthy assassin, the Lion.

"I see another descendant whose foot is a little more strongly hoofed than mine. This change began as a 'sport,' responsive, perhaps, to a change in the animal's habitat. It is of advantage and nature will hold it and build another order of life on it. She will build on it the hoofed Ruminant. Looking far down along this diverging branch I see my posterity in the Ox, in the Lama of Peruvian mountains, and in the Camel of Bactrian deserts.

"Time has wrought other changes. A little animal I see, one of my own descendants, clumsy of body and short of limb like myself, but slightly unlike myself in the foot. The third toe is sheathed in a hoof a little larger than mine. As my descendants are separating from each other in their habitats and modes of life, this one, I see, is disposed to shun contention and seek safety by *running*. Running will beget faster running. The clumsy ankle-joint, by continual flexing, will become more flexible. And those members of the progeny in which it appears *most* flexible, having the best chance to escape the unfolding order of Carnivora, will survive and repeat themselves in their offspring. The structure is put at interest. And as Nature, on this line, is leading on toward greater and greater speed, all parts of the animal frame concerned in locomotion will undergo a modification correlative with that of the ankle-joint. The heel-bone which in me is short and flat, in my remote posterity along that line will be long and slender. The foot which in me is flat and broad and terminated by five toes with impotent hoofs, in that far-off posterity will be terminated by one toe sheathed in a strong hoof, and prolonged and elevated into a lever which shall act in concert with the leverage of the elongated limb to give the greatest speed of locomotion. At the end of that divergent line my posterity will appear in the swift-footed orders of the Deer, the Antelope, and the Horse.

"Another member of my far-off progeny I see, not taking to fighting and flesh-eating, but trying to escape from its fiercer cousins by *climbing*. As yet it differs but little from myself. The 'sport' which made it at first a variety, better adapted for climbing trees, came in a way analogous to that which in far later times shall affect one of Seth Wright's lambs, giving it short and crooked legs and making it the parent of a race badly adapted for jumping fences. It is of advantage and, becoming congenital, it will be emphasized from generation to generation and lead to an arboreal order. Such an order will be fruit-eaters, and as my own quadra-tuberculated teeth would serve well for fruit-eating, the teeth will undergo but little modification. The crowns of the molars will retain the four tubercles.

"The feet, not being brought into use for fighting or running, will undergo but little modification. A slight departure from the primitive pattern will adapt them for climbing. The modifications must affect chiefly the limbs and the brain. The fear and cunning which lead this descendant of mine to attempt climbing will lead in the end to a posterity of cunning, arboreal Apes.

"*Fighting* leads to the murderous orders of the Carnivora. *Grazing* leads to the peaceful tribes of the Ruminant. *Running* leads to the beautiful orders of the Deer, the Antelope, and the Horse. *Climbing* leads to the cunning orders of Lemur, Half-ape, and Ape."

These imaginary previsions are stated broadly. In the most general way they set forth the working of two factors, variation and inheritance. That changes in the animal, responsive to changes in its environment, do occur we have the testimony of every animal imported from one climate into another. That such changes are sometimes very great is attested by facts within the reach of every observer. The Pig lives by grubbing. Its nearest living relative is the Hippopotamus, and if we could trace their lines of ancestry back far enough we would doubtless find them meeting in a common progenitor. But the snout of the Pig has been modified for grubbing and that of the Hippopotamus has not. Now the domestic Pig differs very greatly from the wild Boar. Why? Von Nathusius has shown that where the looseness of cultivated soil reduces the labor of rooting, the skull-growth is arrested and the face is shortened. Now this cause of variation nature herself supplies wherever the wild Pig wanders from the harder soil of a forest to the looser soil of an alluvial bottom. Amphibians extend their range whithersoever a supply of insect-food may invite. Everybody knows that their eggs are laid in water and their life begins as fish-life. But Salamanders have been driven in search of food far up along the sides of mountains, where there was no stagnant water to receive their eggs and start the new generation along its courses of life. What has followed? There being no marsh on the mountain to receive and develop their eggs, they have made a sort of marsh in themselves. The eggs are retained in the egg-duct and fluid is secreted from its walls which answers the purpose of marsh-water. In this fluid the eggs are hatched and in this fluid the young Salamanders disport themselves and live the first cycle of their lives. Behold now a wonder! Behold in the body of the Salamander a tragedy which the whole world, through all the æons of its life-history, has enacted on a scale infinitely grand and terrific! There is hunger. What do the hungry mouths do? Devour the mother in whose body they are housed? They would if they could, for hunger knows no law and nature has no morals. They fall to devouring each other! The older and stronger greedily seize the younger and weaker and ravenously devour them. "Jacob and Esau struggled together in the womb."

Such change in the Salamander has a change of habitat induced. Changes equally great have been induced in the Rabbit by a change of environment. In the year 1419 a few little Rabbits, born on board of a ship, were turned loose on the island of Porto Santo. They increased enormously, for the island had no Carnivora to prey upon them. They became so multitudinous and pestiferous as to drive from the island a colony of men. This Rabbit has become small in size, nocturnal in habits, and rat-like in form. It has departed so widely from the type of its European ancestor that it no longer pairs with it. A new "species" of Rabbit has been created within four hundred and fifty-seven years.

Fluctuations less extreme than these affect every living thing exposed to fluctuations of heat and cold, light and darkness, rain and drought, hunger and satiety. No two men in all the world are alike, because they are not conditioned alike. No two sides of a face are *precisely* alike, because they are not conditioned precisely alike, and do not receive the same blood-flow and nerve-flow. No two blades of grass are alike, because they are not bedded in the same atoms of soil and bathed in the same drops of dew and in the same floods of light. No two leaves are alike, because they are not fed by the same sap, not shaken in the same way by the same wind, not obscured always by the same shadow, not exposed always to the same light and heat. As floods of light and heat from the sun ebb and flow with the seasons, so ebbs and flows the tide of life. As the forces and conditions which environ life fluctuate, so fluctuate the habiliments in which life robes itself. In the nature of things most of these fluctuations, like tidal fluctuations on a shore, arrange themselves about a certain mean. And in the nature of things most of them affect only the individual, or individuals of a few generations only, and the species remains for a long time at the mean or average condition.

So much for the first factor, *variation.*

Wonderful is the tenacity with which nature clings to her havings and gettings. Whatever she gets she holds, whether it is healthful or hurtful. Dr. Brown-Séquard, by surgical operations, induced the disease of epilepsy in certain Guinea-pigs, and found that the disease was transmitted. It became hereditary. Even habits arising from the disease were inherited. Where a surgical operation produced a malformation of the skin, even that was inherited. When a diseased male transmitted through a healthy female his diseased condition to two or three generations of offspring, his mate became, at last, diseased like himself. Facts of terrible significance!

> "Be thou a spirit of health or goblin damned,
> Bring with thee airs from heaven or blasts from hell,
> Be thy intents wicked or charitable,"

nature will receive thee with the same hospitality. But hurtful things must run their course sooner than helpful. Let a variation occur which is helpful to an animal, and in spite of nature's indifference to good and evil it must be perpetuated and intensified. Let us consider two cases affecting our own race.

In a certain part of the valley of the Quillabamba in South America, mosquitos appear in such prodigious multitudes as to make life, to thin-skinned Mammals, almost insupportable. Tribes of men seem to have been exterminated or driven into regions less infested. The few men and women who live there and withstand this cloud of buzzing lances, are found to have thick, mosquito-proof skins. How came this semi-pachydermatous race? Certainly not from a semi-pachydermatous, mosquito-proof Adam, created there on the banks of the Quillabamba. They must have been differentiated from an ancestry with average skins. And this must have been brought about by intensifying and fixing what was at first a mere variation. A man or woman appeared, more mosquito-proof than the others. This person with thickened skin had the advantage over his fellows. His vitality was not drained away by mosquito bites. His offspring were vigorous, and like himself, a little pachydermatous. The *most* pachydermatous among them had the best chance, and so it was in the next generation, and the next, and the next. Thick skins and *thicker* skins were multiplying with each generation until the whole tribe became thick-skinned.

Take another case. The Fuegians of Patagonia, clad in skins of sea-otters or guanicos, and often not clad at all, endure the rigors of the most wretched of climates. They sleep naked on the ground, pelted by driving storms. They dive naked into an ice-cold sea and snatch star-fishes and spiny sea-eggs from the rocks. Unclad women break the ice and dive into the water below with their naked babes strapped to their backs. We look with wonder on such feats of endurance, but we must note that Fuegians are fortified against their wintry sky and sea by a large development of adipose tissue, similar to that which coats the Whale in icy seas of the same latitude. When we see the unclad woman dive into water choked with shards of ice, we must remember that under her skin she wears a thick under-garment of nonconducting fat. How came the blubber-coat on the Fuegians? If it were initial and not derived, we would have to suppose a blubber-coated Adam created on the sea coast of Patagonia. It must have been developed in the same way as the thick skin of the Indian on the Quillabamba. This process has been called by Darwin, "Natural Selection," and by Spencer, "Survival of the Fittest."

And now if the Quillabamba Indian came by his thick skin through Natural Selection, where is our warrant for supposing that the Rhinoceros came by his thicker skin through miracle? And if the Patagonian Indian came by the layer of warmth-retaining tissue under his skin through Natural Selection, by what right of fact or logic can we suppose that the Greenland Whale came by the thick layer of blubber under its skin through a different process?

Usually the process of Natural Selection involves many complications. While the Quillabamba Indian would be getting his thick skin, a reciprocal movement would be going on among the mosquitos. The insect which varied from the average in the direction of sharper and longer bill would have the advantage over average mosquitos. She could get her bill through where they could not. In the same way as the first thick-skinned variety of man was put at interest, so the first long-billed variety of mosquito was put at interest. Whichever would reach the limit of variability first would have to succumb. While the order of Carnivores was unfolding and its members growing fiercer, the Herbivores, in order to survive, must have been growing more wary, and more fecund or more fleet. And while the Herbivores were growing more shy the Carnivores must have been growing more stealthy. And while the Carnivores were becoming more stealthy the senses of the Herbivores must have been growing more acute. Each thing was in correlation with every other thing, as each force with every other force. Independence and isolation are impossible. Carlyle has said that when an Indian in the wilds of Oregon smites his squaw all men and women smart for it — so complex is the net-work of relations between men. Equally complex is the net-work of relations among animals. And as the wits of men are sharpened against the wits of other men, so are the wits and senses of animals sharpened against those of other animals. And as the men and races whose wits are not sharpened by these interactions lapse and take lower rank, so the animals whose structure and senses do not advance responsively to the advancement going on in other organisms about them, must suffer retrogression.

A very suggestive fact is that lately revealed by studies on the brain-cases of the earliest mammals. "Eobasaleus," in English "the King of the Eocene," although as large as an Elephant had a brain not more than one-eighth as large as the Elephant's brain. Uintahtherium, King of the later Eocene, and almost as great in bulk as its predecessor, had a brain but little larger or more complex than a Kangaroo. The Mastodon's brain was small and simple as compared with that of its modern congener, the Elephant. The further we push our inquiries back along the line of Horses the smaller we find their brains. Put *Mammals* for Horses and the statement holds equally true.

The senses of the early Mammals could not have been quick. The vision of Uintahtherium must have been very defective. Its small eyes were overhung by horns and cranial walls so that they could not see upward. They were shaded behind by great bony crests so that they could not see backward. The large muzzle was so in the way that they could see but imperfectly forward.

The limbs of most of the early Mammals were short and their gait stiff. A Carnivore as strong and intelligent as the Tiger would have played sad havoc among a troop of Uintahtheriums and Anchitheriums and Poebrotheriums. On the other hand, these troops, so dull of sense and clumsy of limb, enjoyed a tolerable immunity from such poorly furnished Carnivores as the Hyænodon and the five other carnivorous genera, their cotemporaries.

So the world has advanced, change responsive to change through a vast maze of inter-relations. So the world has advanced through war generated by hunger.

Another factor is love.

Hunger has bred sharpness of claw and tooth, and fleetness of limb. Love has bred beauty, and sometimes strength. Males contend with each other for the female, and thus the sexual passion plays the same part as the hungry stomach. This is one mode. There is another in which no passion appears and no result follows but one of beauty.

Many plants can set no seed unless they are visited by insects. The pollen is carried to the stigma on the bodies of insects which visit the flowers in search of nectar. Now insects are attracted either by odors or colors — moths, in general, by odors, and bees by colors. Suppose that a flower with pistil and anthers so related to each other that it cannot be self-fertilizing and cannot be fertilized by the wind, varies from the average in the way of a brighter corolla or stronger perfume. It will have the advantage over its fellows. It will be more likely to attract the eye or the olfactories of an insect and get its pollen scattered from the anthers to the stigma. It will be more likely to set seed and perpetuate itself. Its brighter corolla will pass, by inheritance, to the next generation and nature will take it and build on it as on an advantageous structure in an animal. The fig which bears its flower inside of its fruit could not set seed without the help of gall-flies which pierce the flower. There could be no figs without gall-flies. And the gall-fly would not pierce the fig unless it had become in some way attractive to it, and useful to it. The rose and great-flowered magnolias must secure the attention of chaffer-beetles. The scarlet-colored flowers with bag-like corollas, so common in the tropics, depend on the visitation of humming birds, which they attract by their brilliancy. The floral world has put on its beautiful robes little by little, because there were eyes; and little by little it has developed its fragrance because there were organs of smell. What the flower does to attract the insect, in the animal kingdom the female does to attract the male, or the male to please the female. The plumage of birds, the beauty of women, and the courage of men, are largely the result of "sexual selection." Natural selection works for the maintenance of the individual; sexual selection — which is also natural — works for the continuance of the race.

So the world has advanced "through hunger and through love."

But all things have not advanced. As among men so among animals there has been advance and then recession. The struggle for life, if too sharp, works degradation. If it drives a Mammal into the ground it will deaden its senses and transform it into a Mole. Moles and Bats are extreme modifications of a common insectivorous ancestor, the Bat being a later divergence, as in embryo it bears a close resemblance to the Mole.

If the battle for life drives an insect into a cave, touch and hearing may be quickened, but sight will be atrophied. Blind Beetles are found in the dark recesses of Mammoth Cave. They belong to two families, the *Silphids* or burying beetles and the *Carabids* or carnivorous beetles. The large Silphids deposit their eggs in dead birds, mice, and the like. The small species live in fungi, in carrion, and in ants' nests. The species found in the cave belong to the lowest members of the family, as if they had lapsed before they took to cave-life. Their habits of life have undergone a complete revolution.

More abundant than the Silphids are the Carabids. They must be reduced to the extremity of preying upon their own young when the supply of Silphids is short, and short it must always be. In these dark grottoes hunger is perpetual. "*Blind and hungry*" may be written of all things whose abode, for many generations, has been the sunless solitudes of a cave. Dr. Packard has shown that the principal variation going on now in cave-beetles is in the way of reduction of size. The factor which brings it about is hunger, overmuch hunger — *rabies ventri.*

If, in the life-struggle, one organism fixes its abode on another, there is hardly a limit to the degradation which awaits it. Fig. 60 represents a something which one would be at a loss how to name if he did not know its life-history. It is a sort of cylindrical mass with a bundle of roots growing from one side. Its form might lead you to classify it as an aborted vegetable growth, but its substance leads you to place it among animals. Senses it has none, limbs none, organs none. It has hardly the *vestiges* of organs. There is not the faintest trace of eyes or limbs, and only the faintest trace of mouth and intestine. If it is an embryo it is not far enough advanced to show us what it is going to be. If it is a degraded maturity it has sunk too low to show us what, in the person of its ancestors, it has been. It is not an embryo, and *its* embryo will tell us whence it came, while its habits of life — existence, rather — will tell us *how* it came. This structureless clump begins life as a little Barnacle-crab swimming through the ocean, having a segmented body, three pairs of legs, and one eye. In this early stage, called the "Nauplius stage," it bears a close resemblance to the embryos of all Crabs. Very soon our Nauplius Barnacle leaves off his roving ways and attaches himself to the soft hinder parts of a distant relative, the Hermit-crab. There he does nothing but absorb the juices of his host. Soon he loses his eye, then his limbs, then the segmentation of his body, his head, his intestine, his everything. He grows too lazy and sinks too low *even to eat.* Around his mouth develops a bundle of roots which spread through the soft body of the Hermit as roots of a plant through the ground. He absorbs nutriment like a plant, and hence *all* the ani-

Fig. 60.

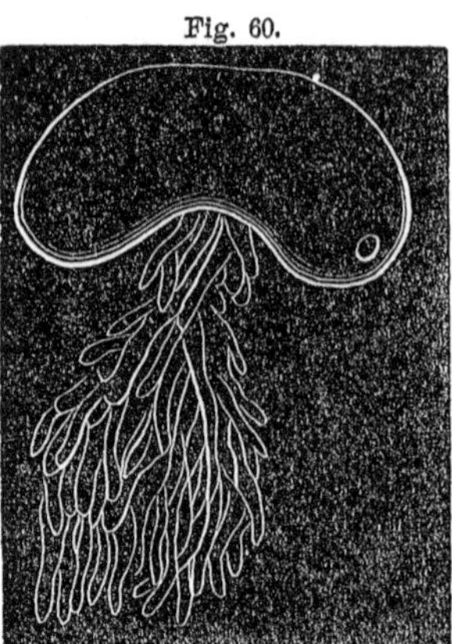

Fig. 60. Sacculina perpuria.

mal structures, even the intestine, aborts, and the purple Sacculina — that is its name — has sunk to the level of a mere unconscious existence. Host and parasite started alike. The Nauplius Hermit might be mistaken for the Nauplius Sacculina. In some long ago, before either Hermit or Sacculina had come to be, the common ancestor of each, the common ancestor of *all* Crabs, wore in maturity the form of this Nauplius. Parasitism has pulled the Sacculina, and a host of other Crustaceans, down into utter debasement.

Retrogression through parasitism has affected a countless host of insects. All creepers on the body, whether of bird or beast or man, all burrowers *in* the body, have lapsed from an ancestry of higher rank. The *cimex* is probably a lapsed hemiptera, and the *pediculus* a lapsed diptera. Lower and lower grades of parasitism are indices of further and further retrogression.

Perhaps the darkest fact on the face of nature is parasitism. Nothing escapes. No organism large enough to be seen by the eyes of men is free from infestations. On this side of nature things are bad, *utterly* bad, *shockingly* bad. A female buries her jaws in the gills of a fish, and hangs there every moment of her life, sucking the life of her host. The male buries his jaws in the body of the female, and hangs there sucking from his mate the life which she is sucking from the fish! Such moral disorder is simply appalling. We stand aghast before the pangs inflicted on all ranks of conscious life, with no compensating good to the victim and no enjoyment to the parasite. To interpret such moral disorder from the old point of view is to introduce it into our own minds. Consider the last attempt. Van Beneden, after describing the frightful sufferings inflicted by the *Lucilia hominivora*, indulges in what are meant for religious reflections. He speaks of the "ever-helping Hand" which provides for the welfare of these repulsive devourers of men. He speaks of "the wisdom" and the "beneficence" which superintends the preservation of parasites which he describes as devouring the body of a Mexican, destroying first his glottis, then the sides and roof of his mouth, rendering them "torn and ragged as if a cutting-punch had been driven through them." And this poor man, whose life is a continuous torture, whose body is the prey of loathsome guests, so hedged in by "the benevolence of Deity" that he cannot destroy them, is exhorted to lift his voice in gratitude and thanksgiving! It is so "ordered" and "contrived" that the frightful Lucilia shall have a brood of offspring — but Newton dies childless! This conception is horrible. The mind that entertains it invites the moral disorder without to reign within.

From the old point of view, the darker side of nature, when regarded as miraculous, is inexplicable. From the new, when regarded as non-miraculous, it is intelligible, and it comports with moral order.

So creation advances and *recedes*. So there is progression and *retrogression*.

And as some orders have risen and others have lapsed, so in the same organism, some parts have advanced while others have remained stationary and others still have receded. Bears have been modified into flesh-eating and fruit-eating kinds, but their teeth have resisted change and remain the same for all species. The little Aye-aye of Madagascar has become almost a monkey although its teeth remain nearly like those of a squirrel. The great Elephant has departed widely from the average Mammal in skull and teeth and trunk but it retains the generalized form of limb, and in some of its structures is closely related to the gnawing Rat. The Bird has advanced wonderfully from its ancestor, the Reptile, but the advance has not been along all the lines of structure. The quadrate bone through which the lower jaw articulates with the skull, the single condyle by which the head articulates with the neck, and the unbranching wind-pipe, are parts which still retain the ancestral pattern. MAN has advanced and become the very "paragon of animals," but the advance has been in head and arm and hand. His digestive system has lagged, his circulatory system has lagged, a great portion of his anatomical structure has lagged.

So the organic world as a unit has advanced and receded, and so a single unit may epitomize the whole and in itself exhibit development and arrest of development, flexibility and rigidity, advance and recession. An individual through its span of a few months or years "is a moving equilibrium," adjusting itself every moment to the changes going on about it; and the organic world through the span of the geologic æons has been a moving equilibrium adjusting itself to the changing aspects of air and land and sea and sun, and adjusting each part to the changes going on in every other. The correlations are so complicated as to baffle our finite intelligence. The great problem of astronomy, the problem of the moon, is simplicity herself as compared with the great problem in biology, the Origin of Species. To interpret all the perturbations of the moon under the pull of the earth and sun and planets and constellations, and to map all the windings of her path on the heaven has baffled the skill of the astronomer. To interpret all the perturbations of the organic world, and to map all the windings of its path along the ages may baffle the powers of any mind which is finite. Nevertheless, the astronomer knows that all the factors of his problem are in the spheres, their ever changing collocation and their reciprocal pull on each other. And the biologist knows that all the factors of his problem are in the light and heat of suns, and in the ever changing face of the earth, and in the correlations between eater and eaten. And as the astronomer may so far decipher the interdependencies of the heavenly bodies and their pull, the one on the other, as to enable him to project his mind into the past and map the face of a constellation as an eye might have seen it before man had appeared, so the geologist has so far deciphered the correlations between tribe and tribe as to enable him to project his mind into a remoter past and restore the leading features of its life.

Asia has led the van of the world's progress. At the dawn of the Mammalian life, although her rocks may not record it, the orders must have commenced their unfoldment in such relations to each other as to present the best system of check and stimulant and counterpoise.

Australia has lagged in the rear. When life dawned on that continent the orders did not unfold advantageously. The Carnivores did not start well and the vegetarians, lacking the stimulant administered by tooth and claw, lagged, and all forms of life lagged and became rigid.

South America received from North America a

Fig. 61.

half-made Camel which she never improved. The Lama remains as she took it and is the highest of her ruminants. She made nothing better of the Pachyderm than a Peccary, but she developed the humble Edentate into the bigness of an Elephant.

As the astronomer inferred from the pertubations

Fig. 61. Megatherium.

of Uranus, the presence of an unseen planet beyond, the geologist infers from the unseemly development of the Edentate, the absence during this unfoldment of large and ferocious Carnivores.

In the period called "Pleistocene" the Edentate culminated in South America in such unseemly forms as the Megatherium and the Glyptodon. Megatherium was a ground Sloth, enormous and unwieldy. Glyptodon was a gigantic Ant-eater, stiffened under the weight of its armor. In Megatherium structure was so far outstripped by bulk that the animal was hardly viable. It crept from tree to tree on its hind feet and elbows, its fore feet having been reduced to mere paws for digging and clasping. They were a yard long but they were not good. In Asia or Africa where the carnivorous orders developed vigorously,

Fig. 62.

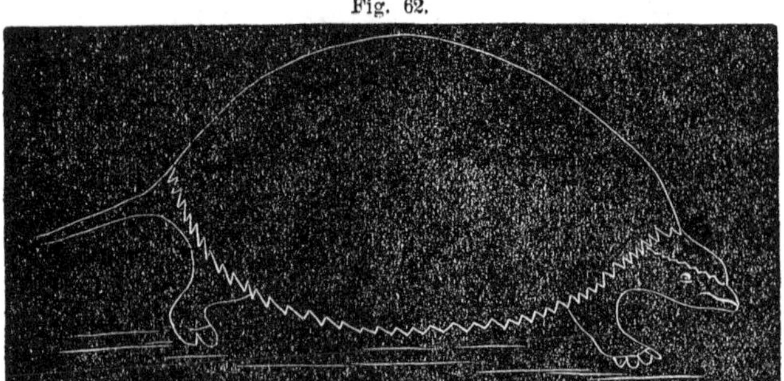

the Edentate order could not have developed a Megatherium. Still less could it have developed a Glyptodon. Edentates develop covering bones. In Glyptodon, through an acceleration of growth, these bones became a huge unwieldy casque not allowing the animal to flex its body. See what followed. The

Fig. 62. Glyptodon.

vertebræ were soldered together and the spinal column became an inflexible rod! Glyptodon became as rigid as if it had carried an iron spike for a backbone and an iron casque riveted to its back for a shield against the bite of ants! Watchful and ravenous enemies, developing at equal pace with the Browsers and Ant-eaters, would have been friends to their posterity by preventing the creation of Megatheriums and Glyptodons.

The elements were not so blended and the orders so balanced in Australia or South America as to lead up to the higher creations.

Where and how did creation lead up to Man?*

* In these pages we do not concern ourselves with the question how life began. The question is one which has engaged the powers of Pouchet, Pasteur, Wyman, Bastian, Tyndall, Huxley, and many others. Recent experiments described at a meeting of the Royal Society by Dr. Bastian, seem to the writer the most interesting and conclusive.

It is well known that neutral or slightly alkaline organic fluids are more prone to undergo fermentation than slightly acid fluids. Chemists have assigned to living germs the property of indicating the process of fermentation.

Wyman demonstrated and Tyndall has again demonstrated, that Bacteria and their germs are killed by exposing them, even for a minute, to the temperature of 212° F.

Bastian has boiled uric acid, or urine, and found that when guarded from contamination it remained pure and barren. He has introduced into this boiled acid, boiled *liquor potassæ*—enough of the potash to neutralize the acid—and found the liquid, in a few hours swarming with Bacteria. These experiments were repeated under many conditions and with every precaution to guard against error.

The conclusions are

1st. *Either the liquor potassæ acts as a fertilizing agent because it contains living germs;* or,

2nd. *The liquor potassæ acts by reviving germs supposed to have been killed in the boiling of the acid;* or,

3rd. *The liquor potassæ initiates chemical changes which result in the production of Bacteria and fermentation.*

In regard to the first hypothesis experiments have shown that boiled liquor potassæ *will only act as a fertilizing agent when it is added in certain proportions.* If it acted merely as a germ-containing medium a single drop of it would fertilize an ounce or more of the acid. But this is never the case.

In regard to the second hypothesis, experiments have shown that a *slight excess of liquid potassæ*, when added to the acid, prevents the development of Bacteria. Moreover, the evidence seems complete that Bacteria and their germs are really killed at a temperature of 212° F.

Now it is shown that the mere development and growth of Bacteria-germs take place in boiled urine containing an excess of liquor potassæ. The inference is that conditions may favor the *development* of organisms which will not allow of their *creation*.

The third hypothesis is the only tenable one. What was once called "spontaneous generation," what is more wisely called *abiogenesis*, takes place in this organic fluid.

One result of the chemical changes through which life is generated is that gases are given off, or being generated, mix with the quickening mother-liquid. In this liquid certain insoluble products make their appearance and reveal themselves as specks of protoplasm. "They emerge gradually," says Bastian, "into the region of the visible and assume the well-known forms of one or another variety of *Bacteria*." "These insoluble particles," Bastian goes on to say, "serve to bridge the narrow gulf between certain kinds of 'living' and 'not-living' matter, and afford a long-sought-for illustration of the transition from chemical to so-called vital combinations."

# CHAPTER VIII.

Origin of Man — Bearings of the Problem — Evolution of the Individual — Evolution of the Race — History of the Brain — History of the Eye — History of the Ear — History of the Mouth — History of the Foot — History of the Hand — The Body of Man a Historic Record — The Animal in him has been Receding — The Man in him has been Advancing — Whence from? — Whither Tending?

WE read in Hindoo fable that the Soors and the Assoors, a race of genii, sat day and night churning the ocean to bring forth the Amreeta, the water of life. Soors sat on one shore hurling the churning staff, and Assoors sat on the other, catching it and hurling it back. They were churning for the water of life, which did not come, but things irrelevant came. The moon was churned out, and sacred elephants and cows.

The fabulist wrote for our own times. Soors and Assoors are not genii, but men, and they churn not the sea, but the sea of thought. Sitting on opposite shores of the sea, they churn to bring forth the *amreeta*, which stood in the mind of the fabulist for the solution of the problem of their own origin and destiny.

Many irrelevant things are churned out, many a white elephant and cow, and many theories, struck of the moon. But the churning goes on and the *amreeta* must come. It is coming.

When Man makes *himself* the object of science,

bating nothing in self respect, he must hold his poise and suffer neither pride nor prejudice to warp his judgment. He must regard himself as from without. He must see himself as he might suppose another intelligence to see him. If he finds a sentiment or quality in himself and then finds a manifestation of the same sentiment or quality in an animal, he must not call it rational in the one and instinctive in the other, or divine in the one and brutish in the other, or good in the one and not good in the other. A noble quality in woman is the love and devotion she bears to her infant. When we see a monkey driving away the flies that pester her infant, shall we call the quality which prompts such an action by another name? A noble quality of heart is that which prompts a man or woman to care for the orphaned and helpless. When we see a female baboon going about with her arms full of little orphan baboons and monkeys whom she tenderly guards, by what name shall we characterize that quality of heart? And shall we say that the affection of a human mother for her offspring is less noble because the animal mother is inspired with the same sentiment? or that human care for the helpless is an attribute less divine because the same attribute dwells here and there in the heart of an animal? Science is Justice with her eyes unbandaged. In her balances attribute weighs against attribute, structure against structure, form against form, humanity against animality, and sentiment does not shake the beam.

Whoever would study man, even in his higher attributes, must study him as if he were a mere animal. He must understand him well in his lower range of passions and appetites. He must understand him in his beginnings, in the unfoldings of his body, and the unfoldings of his mind, and in the gray matter of the brain, where his loftiest thoughts and aspirations are born. Metaphysicians have beaten the air. Hamilton and Mill were men of imperial brains, of the widest range of culture, and of spotless integrity of purpose. Each concerned himself with the highest questions that pertain to the nature and destiny of Man, but with equal ability, equal knowledge, equal industry and equal honesty, they came to opposite conclusions. Must the mighty Soors and Assoors churn forever and get nothing but froth? Is truth unattainable? It is either unattainable or the methods of search have been defective.

Beginnings are alike. It is ends that differ. Turn to page 234, and look again at a representation of the beginning of a Bird and Reptile. Look then at Fig. 63 and see an early stage of the dog. It represents a dog in the fourth week of its development. At this stage you would hardly know it from the forth-coming bird or turtle. And now—

> "Before the little ducts began
> To feed *thy* bones with lime, and ran
> Their course, till thou wert man,"

thou, too, hadst the same form of body, the same club-like limbs, the same reminiscence of aquatic life in gill openings. The plate explains itself. If we ponder it well, it will seem a marvelous fact that the unfolding body of man should traverse the life-phases of all the types that preceded him.

Let us fix our attention first on the history of that organ in which Man transcends all else.

Whatever manifestations of intelligence may appear in a non-vertebrate animal, they have no seat and

Fig. 63.

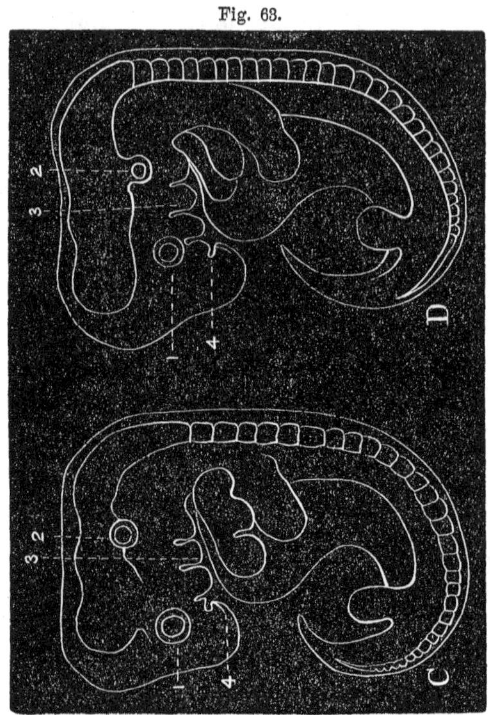


center in a single, specialized organ. No animal below a vertebrate has a true brain.

A brain begins, (Fig. 65, No. 1,) in three little hollow spheres whose peripheries cut into each other, and whose axis are on the same line. The periphery of the third or hindmost spheroid is drawn out in line

Fig. 63. Embryos. C of a Man in the fourth week. D of a Dog in the fourth week. 1, the Eye; 2, the Ear; 3, the Gill arches; 4, the Mouth.

with the axis into a tube continuous with that which encloses the spinal marrow. This is the first draft of a brain. Soon two little bulbs are pushed out from the first or foremost vesicle, the one *a*, upward, and

Fig. 64.

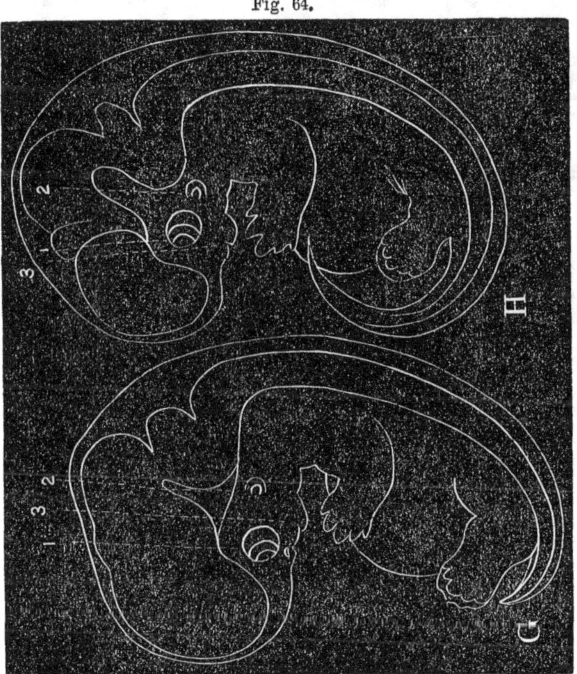


the other *b*, downward, the one called the "pineal gland," and the other "the pituitary body." In this stage the brain appears as at 2. The walls of the three vesicles are of equal thickness throughout. Development goes on and the walls become of unequal thickness. A little bud, *c*, starts up from the third

Fig 64. EMBRYOS. G of a Man in the eighth week, and H of a Dog in the sixth week. 1, the Eye; 2, the Ear; 3, the Gill arches; 4, the Mouth.

vesicle, and another, *d*, forward and downward from the first. A little mound, *e*, rises from the second.

Fig. 65.

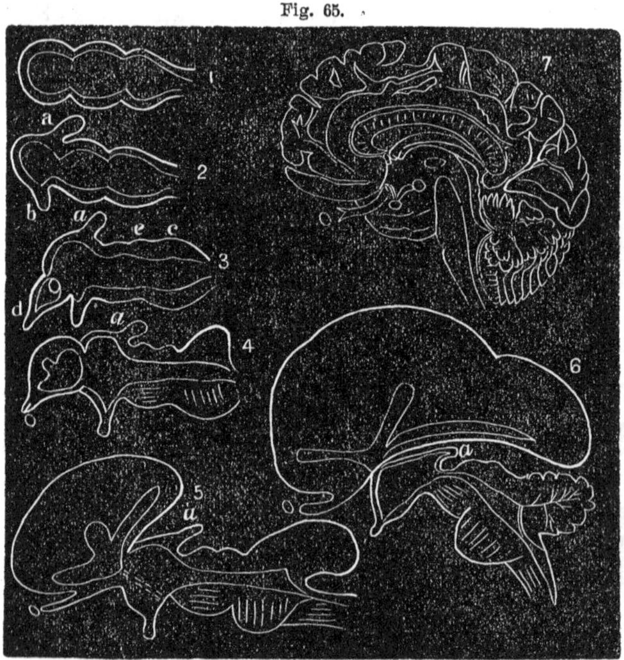


The forward reaching bud from the first vesicle has the front portion prolonged into a lobe. In this stage the brain appears at 3. The bud *c*, from the third vesicle is the beginning of a *cerebellum*. The mound from the second will be the seat of vision. The anterior bud from the first will be the *cerebrum*, the seat of reason. The lobe which it holds will be the *olfactory*, the seat of smell. At this stage the organ of reason is a mere support of the organ of smell, as

Fig. 65. Section of Brain.

a flower-stalk of a flower. From this stage progress is secured by the unequal growth of the parts. The walls of the vesicles continue to thicken unequally, and the "buds" commence to develop now unequally. The anterior bud from the first vesicle grows rapidly and soon becomes as large as the vesicle itself, which has lost its globular form. The globular form of the second vesicle has disappeared in a little tube connecting the cavity of the first, which is called now the *third ventricle*, with the cavity of the third, which becomes now the *fourth ventricle*. The olfactory lobe *o*, has not kept pace, in its growth, with the cerebrum. The brain at this stage is shown at 4. If it were to stop here it would be the brain of a frog. The cavity in the olfactory lobe which, at this stage, would be found even in the human brain, would be retained through life.

Dropping the Reptile here, nature advances still further with her Bird and Mammal. She enlarges still more the cerebrum. It appeared first as a little, downward reaching bud supporting the organ of smell. It appears now as an upward swelling lobe containing a tri-radiate cavity. The olfactory lobe has been left behind, and is now a little, elongated knob, *o*, at the base of the cerebrum. It is evident in what direction nature is moving. She is moving up into organs of *reason*, and is leaving behind the organs of *sense*.

Dropping the Bird at this stage (the stage at which the cerebrum arches back and barely covers the optic mound) creation advances still further with the Mammal. She carries the cerebrum back a little further till it partly overlaps the cerebellum, but she leaves it still smooth on the surface, and at this stage of growth she drops the lower Mammals, such as the Rodents. Advancing from the smooth-brained Rodent, she increases the surface of the cerebrum by foldings and convolutions. She drops now all Mammals except those of a single order, and in this order her advance is marked by the addition of another lobe and by deepening the furrows and multiplying the convolutions. When the little bud which sprouts from the front vesicle has grown upward and arched backward till it overtops the bud from the third vesicle, and rounded out into a convoluted dome crowning the organs of sense, it has reached the utmost limit of unfoldment. Such a brain is that of Man. The primitive vesicles can no longer be recognized. They are masked under their adjuncts. What we saw in an early stage of every brain, what remains the permanent condition of the lowest brain, a little downward-reaching bud from the anterior vesicle, smaller than the organ of smell which it held, that we find now as an overarching dome, crowning and dominating all other brain elements. It began as in the Lamprey, and for a little while Man and Fish traveled on together, alike in brain as well as gill-arches. Leaving the Fish behind, he traveled on for nearly two months with Reptiles and three with Birds, having like them a smooth brain (as at 6) with only one lobe to the cerebrum. Leaving behind first the Reptile and then the Bird, he passed on, companioned only by Mammals. Rodents dropped out of line just as the brain had developed another lobe (as at 6). Other mammalian orders kept pace a little longer, and fell back when the surface of the brain was cut into fissures and crumpled into convolutions. After five months of creating,

Man has sped beyond all the orders of life except that of which himself is head and crown. Another lobe begins then to sprout from the middle or second lobe. This is the posterior lobe, and it comes only on the brain of Man and the higher Apes. Creation advances now along one line, and her steps are registered in a deepening of the fissures and an increasing of the convolutions. Toward the end of the journey Man has but three companions, Gorilla, Chimpanzee, and Orang-Outang. He outstrips them each in bulk and texture and, generally, in the complexity of the convolutions. But such differences as lie between Man and Gorilla lie between one man and another. Our brains are more complex than those of Hottentots. The brain of a Hottentot is more complex than that of a Gorilla. But the *sulci and convolutions which are present in all adult men of all races are present in the brains of Gorilla and Chimpanzee.* Owen failed in his attempt to establish a distinct order for Man, based on the structure of the brain.

Call up, now, the Ape and the Man and see what they will *do* with their brains. The old formula was simple enough. Man reasons and Apes do not; Man uses tools and Apes do not; Man talks and Apes do not; Man worships and Apes do not. But with brains which, to the outward look, are identical in their beginnings, which traverse the same history and are so near of kin in maturity, it would be passing strange if their brain-manifestations were totally and radically different. If Man *reasons* through his cerebrum he reasons very poorly when he infers that the Gorilla must do something entirely different through his.

When Brehm gave an Ape a lump of loaf-sugar wrapped in a paper with a wasp, the Ape, on tearing off the paper, was stung by the wasp. From that day, whenever Brehm gave that Ape, or any other Ape in the cage, a paper package, the animal, before opening it, took the precaution to shake it at his ear and listen to learn whether there was a wasp inside. The Ape had certainly gone through a process of reasoning, and not a very simple process. It involved *generalizing*. The animal must have thought: "Now if one wasp can sting, so can another, and if a man can cheat me once by enfolding a wasp with a lump of sugar, he may try to do it again, and if one man can attempt such a trick, so can another, and if he can attempt it on me, so can he attempt it on my cage-fellows. I will tell them of the trick, and I will tell them to be cautious in their dealings with the biped villains outside." This he must have *thought*, and something like this he must have said. Here was a process of reasoning, a communication of the results, and an exercise of memory.

When a caged Baboon was growing old and his teeth were falling into decay, he conceived the idea of cracking nuts with a stone. His cage-mates saw him and thought the invention was admirable. As there were only two stones in the cage they took these in turn as they could get them, and practiced the new art of cracking nuts *with tools*. The wise old Baboon, whose necessity had been the mother of the invention, tried to protect his rights by hiding the tools in the straw. The other Baboons, in prophetic forecast of Man, watched their opportunity and *stole* them.

These intellectual operations were carried on, not by

the highest Apes, whose brain-convolutions are the same as those of men. No Gorilla has ever been studied in captivity. We know very little of his social life and absolutely nothing of his higher mental operations.

It is safe to say from what we know of the Apes, that the difference between their feelings and our feelings, their thoughts and our thoughts, is one of degree, not of kind. It is safe to infer that in case of the higher Apes as in that of the lower the difference between brains is no adequate measure of the difference between the *working* of the brains. The reasoning of an Ape about the sting of a wasp is the same in kind as the reasoning of Newton about the structure of the universe, but the difference between the instruments is not commensurate with that between the products. Either Newton was vastly greater than the instrument through which he wrought, or the Ape is less than the instrument through which the workings of his mind are carried on. Probably both. The cultivated Man has outgrown his body. The savage man is not up to the measure of his body. In the Museum of the Smithsonian Institution may be seen a cranium of enormous size and most perfect symmetry. Such a noble forehead! and, balanced against this, such a perfect backhead! All the lines and curves so strong, so graceful!

> "A combination and a form, indeed,
> Where every god did set his seal,
> To give the world assurance of a man."

The owner of this head was a miserable Indian who never got from it so much as a beaver trap!

The cultivated races of men so far transcend the

animal that is in them and around them that in the higher range of their faculties they are *super*-natural.

Wordsworth speaks of Nature as

> "The abodes in which self-disturbance
> Hath no place."

He is right, for although nature, in tooth and claw, is red with ravin, yet in one sense hers are the abodes of peace. There is no *self*-disturbance. No remorse dwells in the heart of the tiger. The wolf never condemns herself for killing the lamb. If Man were level with Nature his heart would be as peaceful as the wolf's. But he alone has a sense of the moral disorder which reigns around him and within him.

Leaving now the brain and leaving for a time the attributes manifested through it, we fix our attention on the making of other parts of the body.

The slits, *g*, on the side of the neck, which mark the former position of gills, are dropped, the heart advances from a single pulsating sac into a quadrapartite organ, the feet and hands unfold from limbs which at first are knob-like projections from the body. "The wings and feet of birds," says a great embryologist, "no less than the hands and feet of Man all arise from the same fundamental form." "The great toe," says Prof. Owen, "which forms the fulcrum when standing or walking, is perhaps the most characteristic peculiarity in the human structure." This same toe, so characteristically human in maturity, is characteristically Simian in embryo. Prof. Wyman found it shorter than the others and projecting at an angle from the side of the foot, a form and position which correspond to its permanent condition in the Ape.

As in the growth of the brain, so in the growth of the entire body, Man passes, one by one, the lower orders of life, and it is only in the later stages of development that he passes the Ape. He has passed the higher Ape but very little, even at the period of birth. The young Gorilla and the young human are not very far apart. They grow up, the one into brute-hood and the other into manhood, and every day they grow asunder. In the human the brain grows up and pushes out the skull. In the animal the skull grows thick and pushes in the brain. Acceleration of growth in the human affects the brain and brain-case. In the Gorilla it affects the jaws and nasal cavity. In the

Fig. 66.

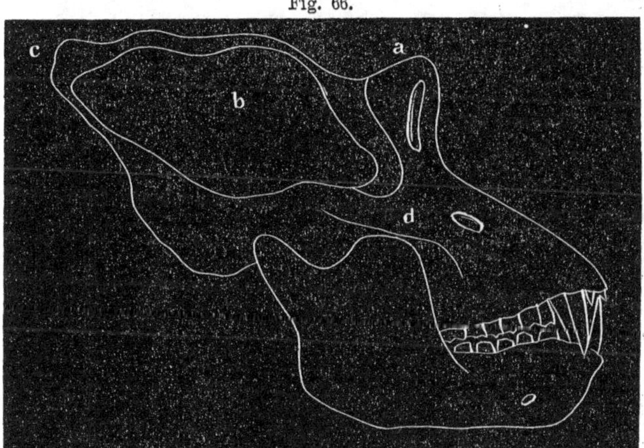


human the arch of the calvaria retains its smoothness. In the Gorilla it develops bony appendages. There comes a great rugged ridge over the eye and a bony crest along the crown which towers like a serrated

Fig. 66. Skull of Gorilla.

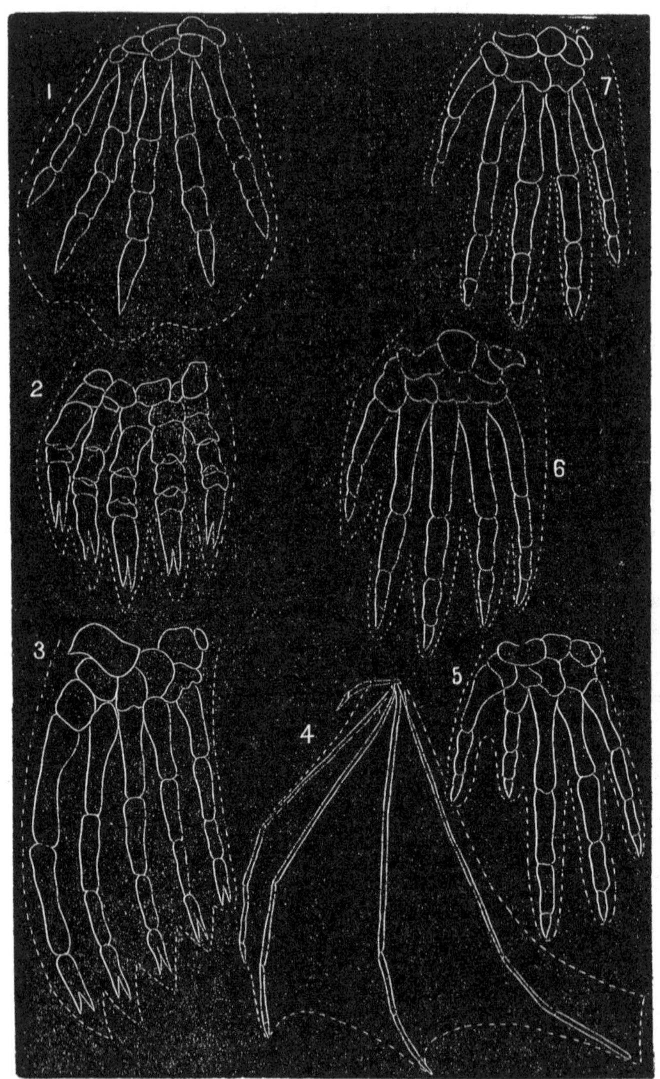


Fig. 67. Hands.

wall above the arch of the calvaria. In the human, creation pushes those parts which have to do with *thinking*. In the Gorilla she pushes those parts which have to do with *biting* and *breathing*. Look at the enormous jaws, then at the enormous nose cavity, and then at the great sinus over the eye, to make way for which the brain has to shrivel and retreat!

The great nose cavity testifies to great powers of respiration. Gorilla must have great lungs and chest. Connected with the respiratory functions is the supra-orbital sinus. A large nose-cavity implies, then, a large sinus, and by so much, a smaller brain.

The human grows *brain-ward*. The Gorilla grows *jaw-ward*. The human grows *sky-ward*. The Gorilla grows *earth-ward*. How far apart do they grow?

Comparing the Man-like Ape with Man, one is impressed by the all-pervading similitude. Differences there are, as in the size of the brain, the form of the head, proportion between the limbs, number of ribs, and generally in the form and finish of the parts, but they are such differences as lie between different men, and different races of men. Gorilla's hand is composed of the same anatomical elements as the hand of Man, and the likeness is carried out even to the finger nails. The fingers are shorter and the palm longer than in Man. But now and then, in the lower races, a man appears with a hand which approximates very near to that of a Gorilla. In one of our museums may be seen the skeleton of a Congo Negro in which the mid-hand or metacarpal bones are very long, almost as long as in the Gorilla. As Gorilla's hand differs from Man's in one direction, that of the Orang differs in an opposite direction, and in this member Man is almost intermediate between the two Man-like apes.

If we concede that these Apes may have derived their hands from a common ancestor, we must believe that Man too came by his hand in the same way, as a modification of the same ancestral pattern. That *all* hands and feet are modifications of one primitive pattern exemplified in the earliest Mammal, a large array of facts constrains us to believe. The facts are luminous in the light of this idea, and no other.

Man is separated from the Australian Duck-bill by the whole height of the column of mammalian life.

This little animal in the totality of its structure, stands nearer the common, primary form of mammals than any other living species. Look now at its hand. The plan is shown in Fig. 67, at No. 1. It consists of three parts, a wrist composed of two cross rows of bones, a mid-hand composed of five long bones, and five digits composed, the first of two bones, and the others of three. All these elements are enclosed in a fleshy web whose form is indicated by the outer line. By its side, No. 2, we place the hand of a Mole. It is composed of the same elements as the first but the bones are shorter and stronger and packed more closely together. In the Mole the hand has become a shovel for digging. In No. 3 we represent the hand of a Seal. The bones in this are longer and more slender, and a fact worthy of note is that while the thumb has become the longest digit it still retains the typical number of bones. It is composed of two bones while the other digits have three. In the Seal the hand has become a sort of fin. No. 4 represents the hand of a Bat. Here are the same bones, placed in the same relations but drawn out, all except those of the first digit, into long and slender rods. In the Bat the hand has become a sort of wing. We represent in No. 5 the hand of a Potto, one of the lowest of the monkeys. In this hand we have the same elements in the same relations, but very strangely, the index finger has become atrophied. It is reduced to a mere vestige, although not a bone has been dropped. In No. 6 we have a representation of the hand of a Gorilla. When clad in flesh it is extremely unlike the primitive hand of the Duck-bill but in the form and disposition of the bones it approaches nearer this pattern than any other of the series. Finally, in No. 7 we have the hand of Man, with not one bone the less, not one the more, not the least change in the number of its elements and not the least change in their disposition! And of all these hands the human is that which, in its osteological structure, approaches nearest the old-fashioned, the undifferentiated hand of the humble Duck-bill! And why should it not? The Mole has made a specialty of digging, the Seal of swimming, the Bat of flying, and the Monkey of grasping. Man has not been a specialist. He alone is *polytechnic*, and while the hand has been modified in the Mole for digging, in the Seal for swimming, and in the Bat for flying, it could not be modified for any *special* purpose in Man without damage to other uses.

If any one will insist that these types are not modifications of one primitive type, but special and independent creations, let him explain the atrophied index in the hand of the Potto. Did the Creator form the Potto directly from the elements, or from "nothing at all," with an atrophied forefinger? Being Purpose itself, why did he create the purposeless? Being Wisdom itself, why did he create the meaningless? And let him explain why organs so unlike as the paddle of a Mole and the wing of a Bat are yet so near alike in their beginnings. Let him explain why structures so nearly identical should be masked under forms so diverse as the paddle of a Mole, the fin of a Seal, the wing of a Bat, and the hand of a Man. Was the Creator so poor in resources?

Fig. 68.

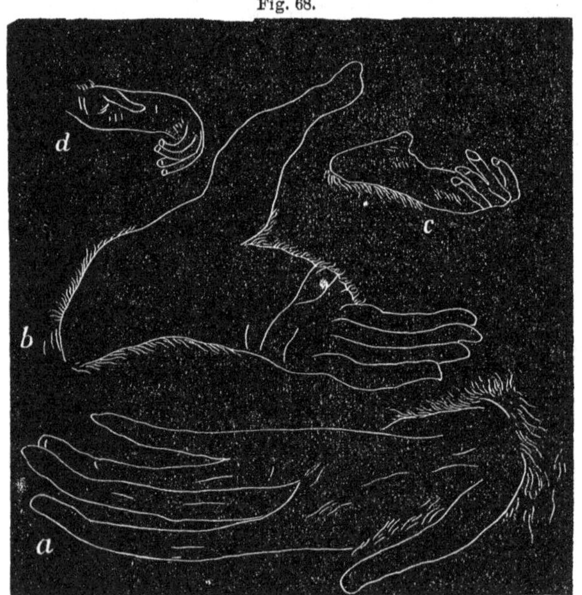


To all these questions science can render a simple answer. Her answer is in two words, *inheritance* and *adaptation*. By inheritance the primitive mam-

Fig. 68. *a*, Hand of Gibbon. *b*, Foot of Gibbon. *c*, Foot of Orang. *d*, Hand of Orang.

malian structure is retained in the embryos of all Mammals, and in the adults of these six classes, and others not figured. By adaptation, the organ based on this structure comes to assume different forms for different uses.*

The history of the foot is quite as significant as that of the hand. The two extremities begin alike. When Man has been in the making only seven weeks you can hardly tell his foot from his hand. If he were born then you might designate his extremities *fore-feet* and *hind*-feet. In the lower Mammal the extremities are all *feet*, fore and hind. In Reptiles and in many Mammals they are all used as organs of locomotion only. Where there is no difference of function we should expect no difference of form. But where the fore-limb begins to take on other functions than those of locomotion we should expect to find it with corresponding modifications.† As we ascend on the scale of life we find the fore limbs relieved more and more from the office of locomotion and given more and more to the service of the head. In Man these limbs are emancipated from the function of locomotion and given over entirely to the service of the mind. They are no longer limbs, but *arms*, and their extremities are no longer feet, but *hands*. Extreme differences of use have resulted in extreme differences of form. The thumb is separated from the fingers. The big toe is not separated from the other toes, but is in line with them. The thumb is opposable, the big toe is not.

* To explain the atrophied index of the Potto we must suppose that this finger was reduced by retardation of growth and that this condition passed down by inheritance.

† See in Fig. 68 the difference between hand and foot, in Gibbon and Orang.

The fingers can be moved, each independently of the others; the toes must move altogether or not at all.

The lowest and oldest order of Mammals, save one, is the Marsupial. The Opossum represents this order in North America. Its habits are partly arboreal and its feet have been modified for climbing. The fore-foot has not been differentiated from the hind, but each has been modified from the *general* pattern. The big toe is opposable. The first step toward a hand has been taken, but it is a step which would lead to the quadrumana, the *four*-handed order. But while the big toe has become a sort of thumb, the other toes have not become fingers. Their movements are still communal. From the Opossum we ascend through the Lemurs or Half-apes to the Apes, finding in all the tribes the same arboreal habits and essentially the same modifications of foot. In the Apes the process of *hand-making* is carried further than in the Opossum, and the toes have become more or less independent in their powers of motion. In the higher Apes, and especially in the Gorilla, the fore-foot is almost as much a hand as in Man, and the hind-foot is left almost as much a foot as in the Opossum.

What relation does the human foot sustain to this history? A relation full of significance. The position of the big toe in embryo, is adapted for prehension. It would seem to be a reminiscence of a remote ancestral toe so formed and placed as to be useful in climbing. On the palm of the hand is a muscle called the *palmaris* which is used in clasping. On the sole of an ape-foot is a corresponding muscle called the *plantaris*, useful in the same way, for clasping and climbing. On the sole of the human foot the same muscle occurs as a vestige, while now and then, and especially among the lower races, it is found large enough to have functions. It would seem to be the reminiscence of a plantaris which, in some remote arboreal ancestor, flexed the sole of the foot as the palmaris flexes now the palm of the hand.

In 1818 Traill dissected one of the higher Apes and found in the region of the thigh a muscle which he thought had no representative in man. He named it the *scansorius*, or "climbing" muscle. Late dissections have shown Traill to have been in error. Its homologue in man is found to be the little muscle called *gluteus minimus*. What is the meaning of this little, useless muscle in man, unless it is the atrophied descendant of a real scansorius?

In that man-like Ape, the Orang, Dr. Barnard, of Cornell, has found a muscle whose homologue has never been found in man. In Orang it occurs as a vestige. It has almost faded out. It occurs in the lower Apes and in the Half-apes, but always as a vestige having no functional value. It appears again in the Opossum, but no longer as a vestige. Thus a muscle which is obsolete in man, almost obsolete in the higher Ape, less aborted in the lower Apes, still less aborted in the Half-apes, is found in the Opossum with its functional value.

These facts are significant. They indicate the lineage of the human foot and limb.

As limbs in their beginnings are alike, so, in their beginnings, teeth are alike. From likeness they grow to unlikeness and separate into molars, canines, and incisors. The molars are modified for grinding, the canines for tearing, and the incisors for cutting.

In the Ape the canines protrude beyond the others and appear in their typical character as flesh-tearers. In the male Gorilla they are developed into great fighting tusks. In Man their office is not fighting but masticating. Their crowns do not project beyond those of the molars or incisors. Nevertheless, they are true canines and their character as such has been clearly demonstrated by Owen, and demonstrated still more clearly by themselves, as they take on, now and then, in the lower races, the character of tusks.

These facts are significant. They indicate the lineage of the human teeth. They show that Man's teeth, like his feet and limbs, were developed in some ancestor into greater differences from the primitive formula, and that they are now in a state of reversion toward that formula.

If the dental system has had a history, so has the digestive, for they are in correlation.

In general, as the food of an animal is more concentrated, the intestine is reduced; as it is less concentrated, the intestine is enlarged. For the less the food is concentrated the greater must be the expanse of surface exposed to the ducts which abstract its nourishment. To secure this greater extent of surface there is a branch from the intestine, called a *cæcum*. In the Marsupials this branch is very long. In the Beaver, whose diet is chips, it is very long. In the Sheep, which lives on anything in the way of grass, it is long. In the Cow, which is a little more fastidious of her grass, it is not so long. In the Carnivores, whose diet is the most concentrated form of nutriment, it is shortest. It would seem that if a species, in the course of its history, were to change, little by little, from less concentrated to more concentrated food, this part of the intestine would contract little by little. Now a shortened part of the intestine is indicated by a little worm-like appendix. This appendix is a closed sac, and is not only useless but injurious. In the Orang it is longer and more convoluted than in Man, indicating a greater reduction of the cæcum. In Man it is variable in size and position. Sometimes it is entirely absent. Sometimes it appears as the merest vestige. Sometimes it is six inches long. Sometimes it is closed through two-thirds of its length. Sometimes it is open through its entire length and closed only at the tip. If it were an essential part of Man it would always be present. If it were not essential, and man's body came directly from the Deity with all its parts adjusted at once to their functions, it would never be present. That it is sometimes present and sometimes absent, sometimes long and sometimes short, sometimes closed and sometimes open, is proof that it is absolutely useless. It is worse than useless. It is hurtful. Little seeds sometimes drop into it, become impacted, and cause inflammation and death.

In the light of comparison the vermiform appendix, though useless, is significant. It registers the reduction of the branch of the intestine from some ancestor whose diet and habits were different from those of man.

THE EAR HAS A HISTORY. Glancing at the plates, (Figs. 63 and 64, pp. 282 and 283,) we will see the ear of these early embryos, Reptile, Bird, Dog and Man, depicted almost exactly as the eye, except that in the first two the eye is larger. In the more advanced embryos, the ear has almost faded out, and in the Bird and Reptile its position only is indicated. In the adult Bird and Reptile no ear is apparent. A little external ear appears in the more advanced mammalian embryos, but you cannot tell, at the stage shown in the plates, which ear is the forthcoming human and which the animal. From this stage the advance is on different lines. When the ear is finished it is usually pointed and moveable in the animal, and is always rounded and almost always immovable in man. The animal "pricks its ears," turns them hither and thither to gather up the sound when it dreads the approach of a foe. Man, or the ancestors of man, used to do the same thing, for he retains the muscle which moves the ear, although he has lost control of it.* The vestige of a point appears in many ears, indicating that the ear in some ancestor was pointed. It is evident that Man's ear, as at present, is not the original edition.

The Mouth has a History. If the reader will turn again to Figs. 63 and 64 he will see that in all the embryos the mouth is late in coming. The embryotic body is far on in its development before the mouth is clearly defined. The eyes have come, the ears have come, heart, stomach, intestines are defined, all the great systems are established, while as yet there is no mouth. He will see also that the mouth does not begin in the right place for a mouth. In a few fishes only, and these of ancient pattern, does the mouth retain through life the position it first assumes in embryo. In the embryos of the Fish, the Reptile, the Bird, the Dog, and the Man, the mouth lies just

* Now and then a man appears who has not lost command of this muscle.

in front of the first pair of gill-arches, and it rises simultaneously with them. In all vertrebrates except Sharks, Rays, and Ganoids, the mouth moves forward and upward from the gills, or the position of the gills. In all vertebrates it opens on the ventral side. Now the ancestors of the vertebrates must have had a nerve-ring surrounding the gullet. The inference is that either the mouth-opening has been changed or the nervous centers have been changed. Dr. Dohrn, a distinguished German naturalist, has framed a strong argument from the belated coming of the mouth, and from its first position, near the gill-clefts, that the mouth which now is is not the mouth which used to be. The present mouth, he argues, existed once and functioned as a gill, while another mouth of the early vertebrate opened on the surface which corresponded to our dorsal surface. Traces of this ancient mouth he finds in the fourth ventricle of the brain, which is remarkable for its great size in the early embryo and its subsequent retrogression.

THE EYE HAS A HISTORY. In all these embryos it appears large, especially large in the Reptile and Bird. But its beginnings are small. It begins alike in all mammals as a little blister on the skin. The upper layer of the skin rises, and the lower sinks into a little pit. From such simple elements and such humble beginnings Creation goes on to form such a wonderful structure as the eye. Its perfections have long excited the admiration of the devout; its *imperfections* begin now to challenge the investigation of the scientific. As an optical instrument the human eye is shown to have seven defects. If we were regarding it as an optical instrument made by a human intelligence we would say that some of the defects were due to imperfections in the material. Matter is but a refractory servant of mind. Man cannot outwork his conceptions into material embodiment without defect or flaw. The geometer may conceive of absolute truth, but if he sets his compass on the earth and measures off his angles and projects his lines, he will find in any triangle he can draw many deflections from truth. Mind and matter are so far at strife that mind cannot fully express itself in terms of matter. Those early Christian thinkers called "Gnostics," saw glimpses of truth when they declared that Deity could not outwork his thoughts and create a material universe, and that He relegated the creation to a Demiurge. The nature of that "*Demiurge*" science, just now, is bringing to light.

"He that formed the eye, shall he not see?"

Yes, and he for whom the eye was formed, with reverence and thanksgiving, shall see *how* it was formed. Seeing defects due to the matter in which "the idea" is expressed, he might say that the eye is the *direct* creation of an intelligence which wrought under the same limitations as our own. Seeing defects inherent in the idea itself he is constrained to say that the eye is not a direct manifestation of the All-wise and All-seeing. Noting the presence of a little structure which is perfectly useless and meaningless, he is led to ascribe to it an historical value. It is the vestige of a structure which has functional value in older eyes than ours.

In the eye of the shark, which represents the oldest class of fishes, we find a third eye-lid, called a "nictitating membrane." This membrane appears in many

Reptiles, which got it from the Fish, and in many Birds, which got it from the Reptile, and in the Duck-bill, which stands near the horizon line where Reptile and Bird and Mammal meet, and in the Marsupials, which were an outgrowth from the order of the Duck-bill. It is not strongly developed in the Marsupial, and in the Ape it has lost its functions and is reduced to a vestige. In Negro and Australian men it is reduced a little more. In the higher races of men it is reduced still more. In Man it appears as a "semi-lunar fold."

Man has received his eyes, as well as his hands and feet and limbs and teeth and ears, by inheritance from a long line of ancestors.

His Body is a Library of Anatomical History.

Many animals have developed a system of muscles by which they twitch the skin. These muscles are especially well developed in such Mammals as cannot scratch themselves — an exercise in which all animals with suitable limbs and feet seem to pass a great part of their leisure. Hoofed animals cannot *scratch*, but they excel in the art of *twitching*.

Remnants of the twitching muscle appears in different parts of the human body. On the forehead they are efficient. On the neck they are well-developed but not efficient. In other parts of the body they are reduced to mere vestiges. They are heirlooms handed down from an ancestor who *scratched* less and *twitched* more than man does to-day.

Every useless part, every hurtful part — and perhaps all that is useless is hurtful — must be regarded as historic. Sir William Gull, the great physician of London, has lately raised a cry against the *tonsils*.

Man, he assures us, would be much better off without tonsils, as useless parts are liable to disease. "Were I to make a man," Sir William is so rash as to say, "I do not think I would put tonsils in him," or, in the same vein he might have gone on to say, "*any of those structures which are relics of a former state of being.*" Sir William would be like a man who would break with the past by burning all the books which record it. More of the past is inscribed in man's body than was written in the burned Alexandrian Library.

Wordsworth sang of the soul:

> "It comes from far, trailing clouds of glory."

Science demonstrates of the body that *it* comes from far. How very far!

We sound the abysses of the past and our plummet brings up from the bed of the Triassic Sea ruins of the first recorded Mammal. They are only teeth but they are enough to show us that this Mammal was a little, insect-eating Marsupial. But as twilight precedes the dawn, so must the Monotrema precede the Marsupial. The first recorded Mammal could not have been the first Mammal. The Duck-bill and the Beaked Mole of Australia and Van Dieman's Land are the only survivals of that most ancient order of Mammals — the *Pro-mammalia* — the Monotrema. The milk-glands of the Duck-bill are marked neither by an elevation as in higher Mammals, nor by a depression as in Marsupials. The milk-glands of the Beaked Mole become depressed at maturity so as to form a little pit into which the nose of the young is inserted. This slight depression of the milk-gland is the first step from the Monotrema to the Marsupial.

If it is better for the little one to insert, not its nose only but its head and even its body, natural selection will, in time, deepen that pit and enlarge it and make of it a Marsupium. We may regard the Beaked Mole as a rigid surviving representative of an early form of Monotrema which had started on the way toward a Marsupial, and the Duck-bill as a rigid survival of another form which had undergone no modification, either toward the Marsupial in depressing the gland, or toward the higher Mammal in elevating it.

The Marsupials, having sprung from the Monotremes, retained a generalized structure. Their organization presents a combination of Reptile and Mammal. The Opossum, in its structure and diet and instincts, is a little of many things. It has a genuine reptilian skull, and it has inherited the low instinct of lying — an instinct found as low on the life-scale as the beetle. It eats all things and lives all manner of lives except aerial. Berries, grapes, fruit, insects, worms, eggs, reptiles, quadrupeds, birds, these are a few of its likings. It idles away the day and prowls about at night. With feet partially modified for climbing, it has an awkward gait in walking. Preferring the borders of streams and ponds, it ambles along from place to place and delights to wade in the shallows.

Man's foot, in its muscular structure, is not very far removed from the Opossum's. In the clasping foot of the Opossum the first step was taken from the Marsupials toward the Apes. We may regard the Opossum as a rigid surviving representative of a tribe of Marsupials which passed, by loss of marsupium, into the Half-apes or Lemurs.

The Lemurs retain the nocturnal habits of Opossums. They lead a solitary, arboreal life, and are restricted to Madagascar, Africa, and Islands of South-

Fig. 69.

ern Asia. They are exceedingly variable in their muscular system. Gradations abound which lead down to the muscles of Marsupials. Muscles which are efficient in the Opossum reappear as vestiges in the Lemur. As the muscles are still variable, so the whole organism has been variable and has diverged into the most aberrant forms. The wonderful Flying Lemur of the Sunda and South Sea Islands is a station on the way from Lemurs to Bats. The short-footed species point the way up from Semi-apes to true Apes.

Of genuine Apes two families were evolved, one with narrow nose, and the other with wide nose and prehensile tail. The first family was evolved in Asia

Fig. 69. Loris gracilis. A Nocturnal Lemur.

and Africa, and was potential of higher unfoldment. The second was evolved in America, and like so many other American types, its development was arrested. Man was to come, not in America, but in Asia, or on some land whose site is covered now by her adjacent seas.

Fig. 70.

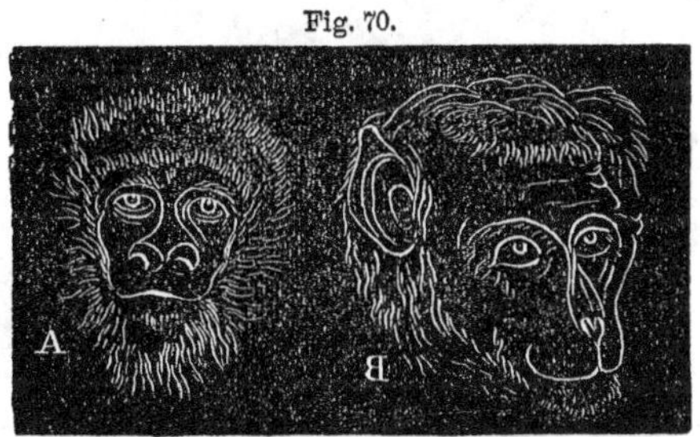


From such Apes as the "Holy-apes" (Fig. 76,) with narrow nose and long tail, the distance was not great to Gibbon and Orang and Chimpanzee and Gorilla. Already the claws had been transformed into nails. Let the tail be aborted, the fore-limbs still further differentiated from the hind ones, and let the commissures of the brain be deepened and multiplied, and nature passes up into the Apes whose bodily structure is the similitude of Man's.

Fig. 70. *A*, New World Monkey. *B*, Old World Monkey. Monkeys of the New World are characterized by wide nostrils and round head. They are called "*Platyrhine*," from platys, broad, and rhin, nose. The nostrils open in front, not below. Monkeys of the Old World are characterized by narrow nostrils which open below, as in man. New World Monkeys have prehensile tails; Old World Monkeys do not. New World Monkeys do not approach men in the form of the jaw or its dentition. Old World Monkeys have jaws and teeth on the same pattern as those of men.

Fig. 71. *A*, Chimpanzee. *B*, Gorilla.

Fig. 72. *C*, Orang. *D*, Gibbon.

THE ANTHROPOID APES, FIGURES 71 AND 72.

Many young Chimpanzees have reached Europe and fallen under the eye of science, but no adult ones.

Previous to the acquisition of permanent teeth the cranium of the Chimpanzee resembles very closely that of a human child. After the second dentition a change takes place and the animal begins to diverge rapidly from the human. A student of the young Chimpanzees brought into Europe, tells us that after the second dentition the forehead sinks below the supra orbital ridge; the jaws expand; the volume of the face enlarges and preponderates over the cranium.

The young Chimpanzee looks like an old, bent, diminutive Negro, but its ways are like those of a playful, trustful, intelligent child. Occasionally it walks upright, but its attitude is generally semi-erect. When tickled it laughs, and when affronted it sulks. Its friendships are strong. It clings with the greatest tenacity to its favorite attendants. When associating with men it becomes fond of dress and will even dress itself. It acquires a liking for tea and coffee and wine and beer, and imitates man by drinking from the cup and feeding itself with a spoon.

On arriving at maturity the Chimpanzee, like all the Simiadæ, undergoes a change for the worse. It is no fault of the animal that this is so, for the price of existence after it has passed from the protection of its parents, is distrust, caution, and perpetual combat with enemies.

Chimpanzees are more terrestrial than arboreal. They climb trees only for food and to command a wide out-look. Travelers who have observed them in their homes say that family ties are strong, that there is a sort of society among them whose basis is the family, and that they build huts nearly in the form in which the natives build their houses. They arm themselves with clubs and fight with great ferocity. They are said to cover their dead with leaves.

Chimpanzees are found in all parts of Africa, from the banks of the Gambia to the kingdom of Congo.

Very little is known of the Gorilla. No live specimen has reached Europe, and competent men have not observed it in Africa. In some features it is more man-like than the Chimpanzee, but it is known to be exceedingly fierce and morose

The Orang, which inhabits Borneo and Sumatra, is one of the

most variable of animals. It varies in anatomical structure, the last joint of the thumb of the hind foot being often suppressed. It varies so much in size of body and form of head that naturalists had founded two or three species where we now know there is but one. The series of skulls which we present in Fig. 73, is from a drawing by Owen, published in Martin's account of the Orang. It will show better than words can the bruteward development of this man-like Ape. It will show acceleration of growth in the facial parts and recession in the cranium. The young Orang with a head like No. 1, is child-like, playful and intelligent. The Orang with a head like No. 2—the same Orang later on in its development—is feebler in brain and stronger in jaw. When the Orang is adult and has a head like No. 3, he is ferocious, morose and melancholy.

The lowest of the man-like Apes is the Gibbon. Its habits are strictly arboreal, and it throws itself with incredible speed and agility from branch to branch. Its brain is simpler than that of the Chimpanzee or Gorilla or Orang, and it manifests less intelligence.

The other Apes, so man-like in their brains, (see page 287,) approach nearer to men in their mental traits than to the common Monkeys. When we see an Orang on ship board climb the highest mast to observe a passing ship, and watch long and intently the fading sails, it is evident that we have here a nobler curiosity than that which impels the chattering Monkey. When we see a Chimpanzee wrap itself in a blanket and go to sleep, or put on pantaloons and cloak, or feed itself with a spoon, or drink coffee from a cup, it is evident that we find here something a little higher than the acts of mimicry we see so often in the cage of a Baboon. When we see a Chimpanzee playing with Monkeys, when ignorant that our eyes are upon it, but never stooping to such society when conscious of our presence, it is evident that we have here something closely akin to personal pride.

Fig. 73.

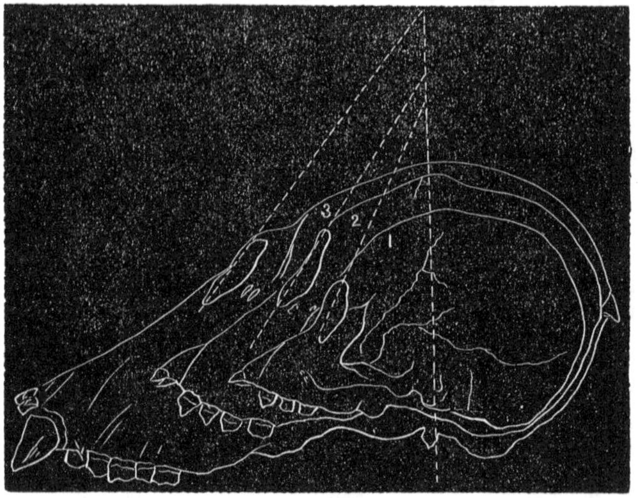


Spoken of the soul we will take the poet's line, just now, on trust. Spoken of the *body* we accept it on proof.

"It comes from far—"

What we have found of chief significance in the history of the parts of the human body is, *advance and then recession.* The limb and foot, from a non-climbing pattern, have been modified for climbing. and then these modifications have declined into mere vestiges. The scansorius and plantaris muscles have been reduced to vestiges, and the prehensile setting of the big toe appears only in the embryo. The canine teeth, from a generalized dentition, have been modified into canine tusks for tearing flesh, and then have lost their tusk-character or retained it only as a vestige. From a generalized muscular system, twitching muscles arose and then became inert and faded out into mere shreds.

The brain contains not a vestige to show that it ever was or could have been in a higher state of unfoldment than now. Its servant, the hand, contains not a vestige to show that it was ever more supple or more polytechnic than now. Its index, the eye, contains a single vestige which shows that once it was not as now, the illuminated window of a soul, but a cold, passionless, nictitating organ, like the eye of an owl.

The animal within us has declined and is declining. The soul within us has advanced and is advancing. A body tenanted by an animal soul will be modified in adaptation to mere animal wants. A body tenanted by a *human* soul will undergo modifications in adaptation to the needs of a higher life.

If an organism varies from the average of its kind in having a stronger jaw, or sharper canine, or larger claw, or more opposable toe or thumb, or swifter limb, or thicker skin, or thicker under-skin of blubber, and if the variation is of advantage to it in its mode of life, nature will avail herself of the structure and go on repeating it and adding to it in the animal's posterity. So arose and advanced the structures which differentiate one animal from another and give it advantage in the struggle for life. If an organism were to vary so far from the average of its race as to have a mind able to contrive advantages for the body, creation will push forward the *intelligence*, and in proportion as that cares for the body, nature will drop her own special furnishings. So arose and advanced the Man, and so declined, in his body, the structures which had equipped the animal. When Man armed his hand with a club, the canine with which nature had armed his mouth, falling into disuse, began to abort. When he stripped the skin from an animal and wrapped it around his own body, the hair wherein nature had clad him began to fade out. When he left off arboreal life, the plantar muscle and prehensile toe which nature had given to his foot, began, the one to abort and the other to undergo the modification which made it a fulcrum to aid in sustaining an upright body.

If the sheep had been able to clothe itself against the cold, nature would not have developed on it an overgarment of wool. If the Arctic Seal had, itself, devised a way of keeping warm, nature would not have developed in it an undergarment of blubber. Now on the Quillabamba in South America, where man has not been able to help himself against the mosquito, nature is working on his body as if he were a mere animal on her own level.* She has made his skin almost mosquito-proof. And in Patagonia where he is still the helpless serf of nature, she is doing for him as for the Seal. She has developed in him a thick under-garment of non-conducting adipose. If the savage were *less* man he would fare better at the hand of nature. If he were *more* man he would fare better at his own hand. Transitions are critical. The Australian, the Bushman, and the Digger Indian are animals and a *little* more. They would be better off if they were animals and *nothing* more.

The vermiform appendix, the coccygeal vertebræ, the carnivorous features of the canine, and the slant of the eye which lingers in the Mongolian race, are

* Marcoy.

heirlooms from a pre-arboreal race. The development of scansorial muscles in the sole of the foot and in the thigh, and the modification of hand and foot for prehension, occurred in a race while becoming arboreal. The reduction of the scansorial muscles and the modification of a prehensile toe into a fulcrum for the support of an erect body, occurred in a post-arboreal race. When the ancestors of the human body were leaving off their arboreal habits, adaptive modifications must have been taking place. The man-like Apes are not strictly arboreal. The structure of their feet implies arboreal habits beyond what the animals possess. The Gorilla, at times, takes an erect position and walks, but not with ease, on feet which are well adapted for prehension. When a race of Apes, owing to changes affecting the forest, or affecting the food-supplies, takes more to living on the ground, it will gain advantage by becoming more strictly quadrupedal or more strictly bipedal. The Baboons have sunk into mere quadrupeds with much of the form and gait of a dog. The Gorillas have risen almost into bipeds with much of the form and a little of the gait of Man. This Man-like Ape is in a sort of transition between a quadruped and a biped.

But Man is the only real biped. The passage from a quadruped to a biped, which some philosophers had thought impossible, could be effected only through a stage of arboreal life which took what had been a quadrupedal body adapted for locomotion on four feet, and left it adapted for locomotion neither on *four* feet nor on *two*. From this intermediate condition, to bring the Baboon *down* was an easy task, as all the ancestral tendencies were downward. To bring the

Man *up* was the crowning achievement of creation. The hands were adapted better and better for prehension, and the feet better and better for locomotion. The pelvis became broad, the spine assumed a peculiar curve, the muscles of the back underwent a correlative change, and the anatomical result was Man. No higher result can follow. When Man had come creation was a spent force. He carries the evidence in certain parts of his body that in him the forming forces were nearing their end. Anatomically, his nose is an unfinished structure. It begins in bone and ends in cartilage, which is simply bone begun and not finished. Prof. Cleland has called attention to the fact that in the highest races the nasal cavity is so elongated that nature can hardly bridge the gap between the crib-like plate through which the nerves of olfaction pass and the palate. There results the most unsymmetrical and ragged bony prominences to be found anywhere in any frame, human or animal. Another irregular and unshapely structure is the breast bone. Another still more unshapely, is the terminal structure of the spinal column — the vestige of a caudal appendage. In the details of the skeleton, generally, we find greater precision of form and neatness of finish in the animal than in Man. His frame represents, already, the ebb of the forming forces. After the human body, *nothing;* after man, nothing but a *better* man. What remains is to work *out* the beast with its appetites, and to work *in* the man with his aspirations.

For the animal which lingers in our bodies, lingers also in our minds.

Animals, if they could speak, might speak to men

in the same vein as the Jew to the Christian: "Hath not an animal organs, dimensions, senses, affections, passions? fed with the same food, hurt with the same weapons, subject to the same diseases, healed by the same means, warmed and cooled by the same winter and summer as a *man* is? If you prick us do we not bleed? If you tickle us do we not laugh? If you poison us do we not die? If you wrong us do we not revenge? If you give us tea will we not sip? If you give us a pipe will we not smoke? If you give us beer will we not get drunk? Are not, then, our nerves of taste the same as yours? After getting drunk are we not, like you, cross and dismal? Is not, then, our nerve system responsive to injury in the same way as yours? If you approach us with a contagious disease do we not catch it? Are we not, then, in tissue and blood as in nerve, similar to you? If you laugh at us are we not chagrined? If you insult us are we not angered? Are we not, then, in the lower ranges of mind, as in blood and tissue and nerve, similar to you? When we provide for the common safety by detaching certain members of the community and posting them as sentries, are we not exercising the same faculties as you, when you appoint your police? Are we not, then, similar to you in higher ranges of mind?"

All these things the animal might say, and more. Crows might urge their claims to mental kinship by telling us how they feed their blind and helpless comrades. Baboons, on seeing the application we have made of the principle of the lever, might urge as a bond of kinship that they too have used a stick as a lever wherewith to pry up stones. And on seeing a band of men go with muffled tread to the plundering of a bank, they might urge as another bond of kinship, that when they go in a band to the plundering of a garden, they too restrain their tongues and tread with noiseless step, and even slap on the face an incautious young fellow who forgets the nature of the enterprise so much as to utter a word of chattering. "If we are like you in the rest we will resemble you in that. The villainy you teach me *I* will execute." Rather, he should say, "the villainy which *I* teach, you have executed, and in this as in all else which has passed from me or my kind into you, you have *bettered* by the instruction."

If the Ape were wise, and anxious to make out his claims to kinship with Man, he might go on in a vein something like this: "You grant a certain kinship in the lower faculties of the mind but you claim powers and attributes of which you say I have not the faintest rudiment. Let me caution you against a fallacy. It is in assuming that what you find in your own mind, exists in the mind of every member of your race. Now the men I see in my native woods are not all like you. Under my Durien-trees of Borneo I see hairy men, somewhat like myself. They are very good in *climbing*, but of what you call 'good and evil' they seem to know nothing at all. If you ask the Gorilla what kind of men he meets under his African trees he will tell you that they are dark of skin like himself; that their nose is flat and hair crispy like his own; that their nature is sluggish; that what you would call the 'moral nature' is rudimental; that what you would call 'the religious nature' fastens on things low and gross. And if you ask the wide-nosed monkeys of South America what they think of man from the specimens under their trees, they will tell you that his range of thought and feeling seems almost as low as their own. When a savant asked a South American Indian what he thought about a life beyond, or a Being invisible, he shook his head. He had never given a thought to such things. And when the great man talked about the starry heavens and tried to rouse the sense of awe and mystery, the savage still shook his head. 'Neither I nor my fathers,' he said, 'ever thought about these things,' and he went on to say something about *hunger*. But *you* have a sense of infinitude. You feel that you sustain relations with the Infinite. You aspire, you adore, you worship. You have the attributes of justice and mercy. Have you ever thought how late in coming were these kingly attributes? What was a court of justice among your ancestors only a few hundred years ago? It condemned to death one of the first physicians of England for the crime of raising a storm by sailing over the sea in a sieve in company with two witches on broom-sticks, and King James of pious memory graced by his presence, the tortures of the execution. Justice! why, there *was* such a sense in the English mind in the times of James the First, and it demanded that Dr. Fithan, of spotless life, for the crime of brewing a storm in a sieve, should be burned for a few minutes by men, and then through the æons of eternity by the Merciful God! Justice — the sense of what is just between man and man and between God and man — so slow in coming, has not yet come into the minds of men in the third sense, the sense of what is just between *man and animal.*

"The injustice and cruelty to animals, so characteristic of the races called civilized, will be held in future ages as one of the crowning vices of a more primitive humanity." All this might be said and more. It might be urged that the virtues first known among men are courage and self-sacrifice, qualities which are not at all uncommon among animals. Even the ant sacrifices itself for the good of the community. It might be urged that the vast majority of men, ever since men began to talk about virtue and vice, have made for themselves fictitious virtues and vices. It has been a virtue to make a long pilgrimage in the attitude of a measuring-worm; to walk with pebbles in the shoes; to stand with arm outstretched and motionless till it grew rigid; to stand in hunger and filth on lonely pillars till the body became a prey to vermin; to wander houseless over the desert; to house in caves with the beast; to flagellate the skin with whips and the stomach by keeping fasts.

Religion, another ennobling sentiment, manifests itself, first, in superstitions. There are savages so near the level of nature they have no religion at all. There are others so little above nature that their only religion is a torturing dread of the Unknown.

After all, the highest animal, if illuminated to perceive what is in its own mind and in the minds of men, would have to concede that it never said to itself, "I ought;" that it never thought of an act as virtuous or vicious; that it never conceived of relations between itself and any Power or Intelligence unknown to it; and that as to Man it is better he should say, "I ought," when he ought *not*, than never to say it at all; and better he should be tortured by a dread of imaginary Powers invisible to him than to have no conception at all of Powers above or beyond.

> "For who would lose, though full of pain,
> This intellectual being?"

Pain is placed evermore at the portals of birth. Through war of each against all, animal species are born. Through war in man of the spiritual against the animal, the higher man is born. When man looks nature steadily in the face he pronounces her a scene of moral disorder. When he looks in on himself he finds disorder there as well, but it does not type the disorder without. Whence came into him the sense of justice when there is no justice in nature? or pity, when there is no natural pity? or mercy, when nature is not merciful? or whence came into his mind that dreadful word OUGHT, which nature has nowhere syllabled in the animal mind? If he pronounce this a scene of moral *disorder*, his mind must picture over against it an ideal state of order. Whence came to him aspiration for such an ideal? If he hunger there is meat. If his soul hunger for righteousness, is there not *righteousness?* To this upward looking of Man there must be an answering fact. And in this upward looking and upward striving, man is ennobled. All earnest and thoughtful men feel keenly the duality within them and strive for the good beyond them. As—

> "An infant crying in the night,
> An infant crying for the light,"

so has been the cry of the noblest souls, out of darkness for light. "As the hart panteth after the waterbrooks, so panteth my soul after thee." "Create in me a clean heart." "Behold, thou desirest truth in the *inward* parts." "I find a law that when I would do good, evil is present with me. For I delight in the law of God after the inward man: but I see another law of my members warring against the law of my mind and bringing me into captivity to the law of sin. O! wretched man that I am."

In words like these, earnest souls have voiced and will long continue to voice their aspirations and their sorrow. For the noblest brows have been wreathed in cypress. He who has seen deepest into nature and life has seen most to oppress and sadden. He who has the deepest sense of the fitness of things is most hurt by the prevalent *unfitness* of things.

Hurt souls will find relief in Darwinism, and the religious sentiment will hold the citadel of faith against the darkest army of facts. When Sharks, and Tigers, and Tasmanian Devils are regarded as no more miraculous than cyclones and tornadoes, their presence on the earth as the seeming embodiments of evil will not wound the mind as now. A larger faith will be born; faith in man, whose beginnings were in the infinite past and who is still in process of creation, and is moving on toward the crowning race—

"Of those that eye to eye shall look
On knowledge; under whose command
Is earth and earth's, and in whose hand
Is nature like an open book."

Faith in God—

"That God which ever lives and loves,
One God, one law, one element,
And one far off divine event,
To which the whole creation moves."

# CHAPTER IX.

Origin of the Races — Branches of our Family Tree — Early Branches from which they Sprang — The Aryan, the Semitic, the Mongolian, the Negritto, Straight-haired Races, Curly-haired Races, Tuft-haired Races, Fleece-haired Races, Hypothetical Race which would Generalize all these — The Primeval Race — Humanity emerging from Brutedom — Survival of Primitive Features in the Negro and in the Mongolian — Process by which these Features were eliminated in the Aryan Trunk and its Branches — Man throwing off the Animal Mind as well as the Animal Body — Evolution of Civilized Races — Barriers — Poverty of Man — Poverty of Nature in Australia, in South Africa, in the Pacific Islands, in America — Man's Development arrested everywhere except in two Centers — Aborted Growths — For Savage Races no Hope — Races which have the Promise of the Future — The Race of the Future.

WE have a saying which passes current in social small talk. It is that human nature is the same the world over. Human nature is not the same over all the world. The blubber-fed Eskimo is not at all like the rice-fed Chinaman. "As far as the east is from the west," so far, in tastes and aptitudes is the Oriental man from the Occidental.

The English Layard sent a letter to a Turkish Cadi making inquiries about the population and industries of a certain ancient village. Here is the Cadi's answer:

"*My Dear Friend and Joy of my Liver:* The thing you want to know is difficult and useless. I have never counted the houses or numbered the men, and as to what this man loads on his camel or that man hides in his tent, that is no concern of mine. But above all, as to the previous history of this city, God only knows the amount of dirt and confusion the infidels may have eaten before the coming of the sword of Islam! It were unprofitable for us to inquire into it. O my Soul! O my Lamb! seek not for the things which concern thee not. Of a truth thou hast spoken many words, and after the fashion of thy people thou hast wandered much from place to place. Will much knowledge create in thee a double belly? or wilt thou seek Paradise with thine eyes? Listen, O my Son! there is no wisdom like belief in God. He created the world, and shall we seek to penetrate the mystery of creation? Shall we say, 'Lo this star spinneth around that star?' Let it spin! Or shall we say, 'Behold that star with a tail goeth and cometh in so many days?' Let it go and come. The hand that held it *can* hold it. O my Lamb! surely thine hour will come.

"The meek in Spirit.

"IMAUM ALI TADE."

In the questioning, time-exploring, space-exploring Layard, occidental human nature spoke. In the lazy, dreamy, care-nothing Turk, oriental human nature spoke.

Human nature is not the same over all the world, but the races differ as their antecedents and surroundings. How were these differences brought about? We answer in a general way, that the *races* of man came through the same factors as the race, and that race-making is simply a later stage of man-making.

We have traced the Mammals back to a generalized Mammal, a Mammal from which divergent lines have led to the different orders. Our method shall be the same with the races of men as with the orders of Mammalia. But while animals reveal their history only through their structure, men have put the last segment of their history into books, and they reveal another segment through their languages, their customs, and their superstitions.

If we represent the races under the symbol of a tree, the Italians, the French, the Spaniards, and the Portuguese, would appear as branchlets growing from the same bough. That bough is the Roman. We glance down along the Roman bough till we find it springing from another bough much lower on the tree. This is the Aryan. We trace the German, the English, the Irish; the Greek, the Persian, and the Hindoo, down to the same bough. They are all offshoots of the Aryan.

South of the Baltic is the remnant of one of the oldest races in Europe. We call these people Esthonians. Their place on the family tree is represented at *b*, Fig. 74. It will be seen that the Esthonians are an early offshoot from the Aryan. They are probably tho first Aryans who found their way into Europe. They are a people of arrested civilization. The Germans who in 1212 invaded their country, found them abandoned to nomadic life and with no hold on the past, not even by tradition. They had no idea as to who they were or whence they came. They remain very much as they were when they first came under the eye of history. They are ignorant and superstitious and bitterly opposed to such innovations as chimneys and windows and knives and forks and light colored bread.

The Esthonians are related to the Finns, who seem

Fig. 74.

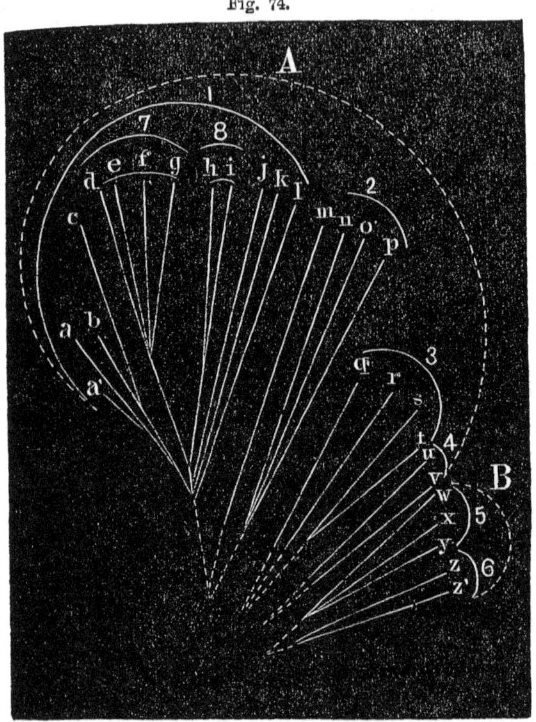


Fig. 74. SCHEME OF THE RACES.—Dotted line *A* embraces the straight and curly-haired Races. Dotted line *B*, the woolly-haired Races. Under line *A*, the unbroken line 1 embraces the Indo-Germanic; 2 the Semitic; 3 the Mongolian; 4, the Malayan. Under the line B, the unbroken line, 5 embraces the Tuft-haired Races; 6 the Fleece-haired Races; *a* the Finns; *a'* the Lapps; *b* the Esthonians; *c* the Hungarians; *d* the Italians; *e* the French; *f* the Spanish; *g* the Portuguese; *h* the Germans; *i* the English; *j* the Greeks; *k* the Hindoos; *l* the Persians; *m*, the Turks, who do not fall under either of the lines 1, 2 or 3; *n* the Jew; *o* the Arab; *p* the Aramean; *q* the Dravidas; *r* the Chinese; *s* the American Indian; *t* the Eskimos; *u* the Malays; *v* the Australians; *w* the Papuan; *x* the Hottentot; *y* the Bushman; *z* the Kaffre; *z'* the Negro.

to have been an early offshoot from their branch, and to the Hungarians, a later offshoot.

More remotely allied to these people are the Turks. From the prairies of Asia they have spread to the shores of the icy sea and to the Mediterranean and Caspian, preserving in all their migrations, their strong race characters. Lamartine has described the Turk in describing his home-country: "This basin, which extends, uncultivated, from the frontiers of China to Thibet, and from the extremity of Thibet to the Caspian sea, produces, since the known origin of the world, nothing but men and flocks. It is the largest pasture-field the globe has spread beneath the foot of man, to multiply the milk which quenches his thirst, the ox which feeds him, the horse which carries him, the camel which follows him bearing his family and his tent, and the sheep which clothes him with its fleece. Not a tree is to be seen there to cast its shade upon the earth, or supply a covert for fierce or noxious animals. Grass is the sole vegetation. Nourished by a soil without stones, like the slimy bottom of some ocean emptied by a cataclysm, watered by the oozings of the Alps of Thibet, the loftiest summits of Asia, preserved during the long winters by a carpet of snow, propitious to vegetation; warmed in spring by a sun without a cloud; sustained by a cool temperature that never mounts to the height of parching, grass finds there its natural climate. It supplies the place of all other plants, all other fruits, all other crops. It attracted thither the ruminants, and the ruminants attracted man. They feed, they fatten, they give their milk, they grow their hair, their fur, or their wool for their masters. After death they bequeath their skin for his uses. Man, in such countries, needs no cultivation to give him food and drink, nor fixed dwellings, nor fields inclosed and divided. The immeasurable spaces over which he is obliged to follow the peregrinations of his moving property, lead him in its train. He takes with him but his tent, which is carried from steppe to steppe, according as the grass is browsed upon a certain zone around him, or he harnesses his ox to his leather-covered wagon, the movable mansion of his family."

More closely related to the Turk in habits than in blood is the Arab, and from the same branch as the Arab is the Jew.

We cannot trace the Aryan, the Turk, and the Semitic races back to a common ancestry. There is a convergence of these races toward a common point of unity, but the lines do not meet within the horizon of history. We carry the branches down by dotted lines till they meet in a common bough or trunk.

Dispersed over both Americas is a race, divided into many tribes which differ from each other, north and south, east and west, under mountains or on plains. But the differences are such as a few thousand years of climate or modes of life might bring about, and the American Indians, north and south, east and west, prairie Indians, mountain Indians, village Indians, and roving Indians, must all be placed, as Morton placed them years ago, as *one race.* They were created not in America but in Asia or some land adjacent to Asia, for no Narrow-nosed Ape is now or ever has been a native of either America. But they were not made out of a Jew or a Turk or an Aryan. They are a branch from some other bough or trunk.

In India there is a race living among the Hindoo Aryans but not of them. The Dravidas have attained only to the rudiments of a civilization, and in their domestic life they present affinities with the American Indian. They are a sort of *generalized* race since they show traits of relationship to the Mongolians on the one hand and to the Malays and Australians on the other. We infer that they are an ancient race. There are indications that although restricted now to a few tribes in the Deccan, in early times they occupied the whole of Hindustan. We draw a line from the Dravidian, convergent toward that drawn from the American Indian.

An ancient and strongly marked race is that of the Mongolian. In early times it separated into two branches, a monosyllabic and a polysyllabic branch. The Chinese, Siamese, Burmese and Thibetans speak in monosyllables. The Japanese, the Tartars, the Kalmucks, and the Tungusians speak in polysyllables. No Mongolian has a long head, and some of the sub-races — as the Kalmucks — have very short heads. They have narrow, slanting eyes, prominent cheek bones, broad noses, and oval faces.

A late outgrowth from the Mongolian branch is the Polar Man. Driven into hyperborean regions this Mongol has found the conditions of life so hard that he has been modified into a race, seemingly as remote from his brother, the Chinaman, as from his cousin, the Jew.

Akin to the Mongolian is the Malay. This race is pelagic. We find it in two branches, one inhabiting the Sunda and Phillipine Islands and Malacca, and the other dispersed over the islands of the Pacific from the Sandwich to the Marianne, and from the Manganeva Archipelago to New Zealand.

The Malayan is another generalized or synthetic race. It resembles the Mongolian and also the Caucasian. In the form of the body the Malay approaches the Mongolian; in the form of the face he approaches the Caucasian.

The Australian is an older Malayan. This is one of the very lowest races on the face of the globe. The bones of the Australian are small and weak, his legs are without calves, his lips protrude, his teeth slant forward and his forehead slants backward.

A very different race is that of the Papuan. The Papuan's forehead is low, his nose large, his lips thick and his hair woolly. It grows, not as the woolly hair of the Kaffre and the true Negro, in an even fleece, but in tufts which twist spirally, and, attaining the length of a foot or more, stand out from the head. The Papuan lives in New Guinea, New Caledonia and the New Hebrides. Remnants of this race are found in Malacca and in the Philippine Islands. The Tasmanian race, so lately deceased, was Papuan. The Papuan is an ancient and dying race.

A race related to the Papuan by the screw-like tufts of woolly hair, is that of the Hottentot. Hottentots, like Papuans, exist as a remnant. They are represented now in only two tribes, and these are in the article of death.

A near neighbor to the Hottentot, but of a different race is the Kaffre. A Kaffre's woolly hair grows as an even fleece over the head, his face is long and narrow, his forehead high, his nose prominent, but his lips are not so prominent as those of the Hottentot or those of another neighboring race, that of the Genuine Negro.

The forehead of a Negro is flat and low, his nose broad and flat, his lips thick, his arms long, his legs short and without calves. His range is restricted to a little belt between the Equator and the Tropic of Capricorn. A very small portion of Africa is appropriated by the Negro proper.

This sketch has been purposely brief. It has been full enough to show that all the civilized and progressive races have sprung from the extinct Aryan race; that most of the arrested civilizations are Mongolian and Semitic; that most of the savage races are the non-Aryans of India, the pelagic races, and, generally, the races of dark skin and woolly hair. The Esthonian race might blend, toward the roots, with the Mongolian, and the Mongolian with the Turk. The race of American Indians unites, at the root, with the savage races of Asia, but the Chinese does not unite with the Semitic, nor the Semitic with the Aryan. When we push our inquiries back to the furthest limit of history, or the limit to which language will take us, we find the Mongol Chinese, the Semite, and the Aryan, standing out as distinct types of humanity. We find another race already very strongly marked. The monuments of Egypt show that the Negro was a Negro fifteen hundred years before the Christian era. He stands out in his dark isolation with a mysterious past behind him, a perpetual challenge to the dogma of unity. An eminent savan expresses the concurrent opinion of all unbiassed naturalists when he says that if Negroes and Caucasians were snails, zoologists would universally agree that they represented two

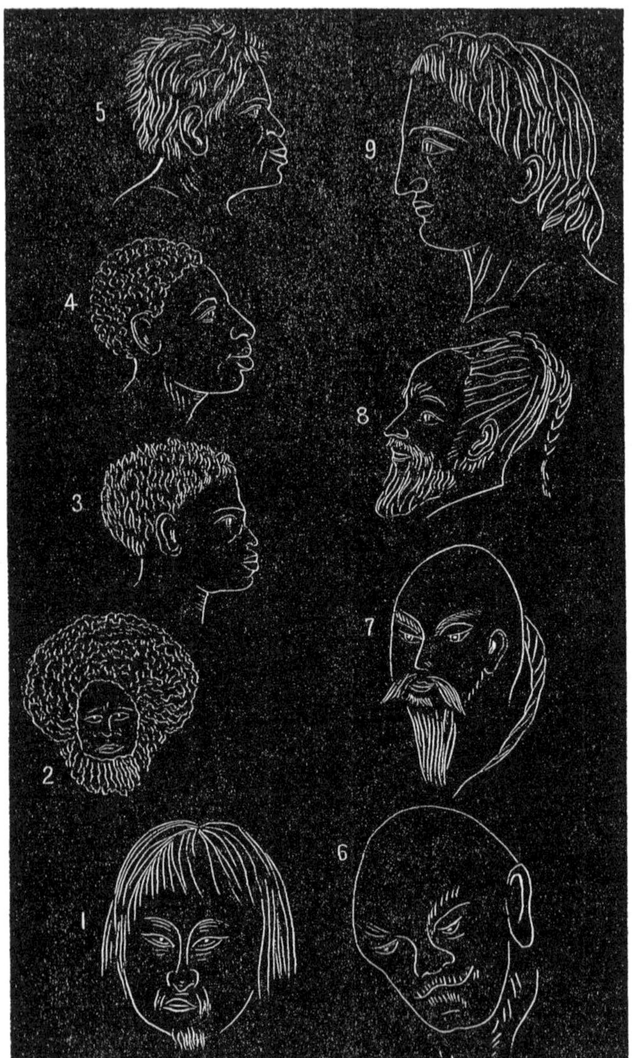


Fig. 75. 1, Eskimo. 2, Papuan. 3, Hottentot. 4, Negro. 5, Australian. 6, "Wild" Chinaman. 7, Chinaman. 8, Arab. 9, Greek.

species which could never have originated, by gradual divergence, from one pair.

We have found certain races which are more or less generalized. The Malay has the body of a Mongol and the oval face of a Dravida man. The Dravida man is intercalary between the Australian and the Mongolian.

Looking at the races from another point of view we may divide them into two classes.

Hardly a feature would seem more insignificant than the hair, and yet no feature is more significant. The Ethnologist is sometimes met with a sneer because he bases a classification on the form of a hair, whether in a cross-section it is circular or oval. He may reply that other features are variable and not always hereditary, but that the form of a single hair on the head, whether it is cylindrical or tape-like, is always transmitted with the blood, is always hereditary within the race. He might reply still further that no race, a cross-section of whose hair is *oval*, has ever attained to civilization, has ever written its language, has ever risen much above the animal.

Tape-like hair is woolly. Cylindrical hair is straight or curly.

The seat of the straight-haired races is the North Hemisphere; of the woolly-haired races, the South Hemisphere. It is only in Africa that the woolly-haired man comes north of the equator.

Of the woolly-haired races there are two branches, the tuft-haired and the fleece-haired

The Papuans, so widely separated from the Hottentots geographically and by language, agree with them in having tufted wool. As the form of the hair is hereditary we infer that Papuans, Hottentots and Bushmen sprang from the same root.

The Australians, widely separated from the true Negroes geographically, by the form of the hair, and by the size and structure of the bones, agree with them in the absence of calves. The Australian is the lowest of the straight-haired races. The Negro is the lowest of woolly-haired races.

While in the straight-haired class, there are races somewhat intermediate between others, the straight-haired touches the woolly-haired at a single point. The lowest of the one meets the lowest of the other in a single feature, *the absence of calves.* Monkeys do not have calves, neither do Negroes, neither do Australians. The Australians are fast dying out; so are the Negroes. Humanity is dying at the roots, not at the top.

All the woolly-haired races, except the Kaffre, bear the characters of age and decay. The Negro race is a very old one, and represents humanity very near its woolly-haired beginnings. The Australian race is an old one, and represents humanity very near its straight-haired beginnings. As both Australian and Negro have dark skin, slanting teeth, receding forehead, broad nose, protruding lips, and undeveloped calves, we infer that if the straight-haired and woolly-haired races sprang from one stalk, such ancestral stalk must have been broad-nosed, thick-lipped, dark-skinned, and without calves on their legs. Was there such a race? Was it human?

We can make out other features of the hypothetical, the generalized race.

It is a well-known fact in biology that a species

widely dispersed is old. The same law should hold as to features. One feature of the Mongolian race is the narrow, slanting eye. This feature is strongly marked in the Chinaman. It reappears with some emphasis in certain tribes of South American Indians. Dispersed so widely, it must be old. We have other evidence than the application of a law. The faces depicted on the walls of Egyptian tombs and temples show the narrow, down-slanting eye. We must infer that this feature belonged to the ancient Egyptian. Faces painted on the palace-walls of Nineveh show the same feature. We must infer that the ancient Ninevite had a narrow, slanting eye. Here, then, is a feature, distinctly Mongolian, appearing long ago in a Semitic race. The Greek, before Phidias, represented the face with slanting, cat-like eyes. As this representation was common in archaic Greek art, we may infer that the Greek, when just becoming a Greek, had narrow, slanting eyes. This Mongolian feature appeared, then, in pre-historic times, in an Aryan race. Chinaman, American Indian, Egyptian, Assyrian, and Greek must have come by this feature through inheritance. To our inventory of features belonging to the pre-woolly-haired and pre-straight-haired race we must add narrow, slanting eyes. The inventory will then read, *undeveloped calves, dark skin, broad nose, thick lips, backward-slanting forehead, forward-slanting teeth, and down-slanting eyes.*

Now the slanting eye is not a feature which could be induced in man by the action of climate. It is a feature of the carnivorous cats, and it has relations to the mode of eating. All animals which have slanting eyes *munch their food.* Planting their jaws in the flesh of a victim, they look while they munch. The cat brings her eyes down to the mouse she is munching, and it is easy to see that such a mode of looking, attendant on such a mode of eating, would result in giving the eye a slant inward and downward. The ancient men, and the modern men of ancient eyes, did not come by this cat-like slope through inheritance from arboreal apes, for being vegetarians, the apes do not munch. They inherited this turn of the eye, as all men have inherited the carnivorous features of the canines, from a pre-arboreal race of carnivores.

The habit of munching implies a flat nose. We may assume that the remote ancestor of all the races had a nose which did not stand out on the face as a feature. The nose — the *human* nose — seems to have been the last created feature. It seems to have come after the formative forces were well nigh expended, for anatomically it is an *unfinished* feature.

Suppose, now, a race of beings — we cannot say men — with dark skins, undeveloped calves, retreating forehead, protruding mouth, flat nose, and slanting eyes. From such a trunk all the boughs and branches of humanity have sprung. Suppose, now, that such a race were to leave off the habit of munching and were to pass into the habit of taking food in their hands — or paws — and lifting it to the mouth in morsels. The mouth would begin to undergo a modification. The lips would be used less as helps in eating. As the mouth and lips would contract, the nose, being in muscular connection with the lips, would undergo a correlative modification. As the external nose is of no use except when modified for prehension, the course of its development would depend on the parts connected with it, on acceleration or retardation of growth, and on the sense of beauty developed in the mind. It is hard to account for the ridiculous nose of the Nose-monkey of Borneo in any other way than by supposing that it came first to an individual through acceleration of growth and afterward passed into the race through inheritance.

Fig. 76.

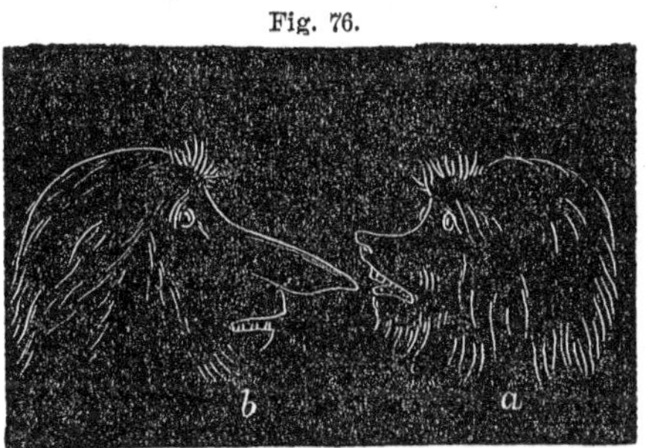


Mouth, lips, and nose changing, the expression of the entire face would change. As mind and brain were growing, the forehead would grow. Its backward slope would finally disappear, and as the frontal would be pushed out the face-parts would be carried in. The mouth would cease to be the prominent feature, the point of convergence for the lines of the forehead and face. If all parts of the body, during this transition period, were plastic, they would share in the modifications undergone by the mouth and nose. Even the eyes, although in no muscle or nerve connection with mouth or nose, would mirror the developed soul and cease their downward slant. Now the primitive Greek, if we can trust archaic art, had a

Fig. 76. Nose, or Holy-Monkey. *a*, the young; *b*, the adult.

protruding mouth, a retreating forehead, a flat nose, and a slanting eye. As a savage, the Greek had a face full of the animal. But his body was supple and plastic and when he became a philosopher he had a face completely de-animalized and nobly human. The Negro failed to develop calves on his legs, failed to outgrow the flatness of nose, the thickness of lips, and the scantiness of forehead which characterized the "common ancestor," and by so much he failed to become a complete man.

The Australian failed in the matter of calves and mouth and nose and brain. The Chinaman and South American Indian failed in the matter of eyes.

The slanting eye of the Mongol, the flat nose, thick lips, and retreating forehead of the Negro, must be regarded as we regard the coccygeal vertebræ, the plantaris muscle and the like, as vestiges of an animal ancestry.

The first important step toward man was that taken by something which climbed a tree. The next was taken by some arboreal, ape-like thing, which came down from the trees and ceased to be arboreal. The next was taken by something which left off munching and began to take its food with the hand, in morsels. For as the limbs were primarily mere organs of locomotion, so at first the function of the mouth was merely to eat. And as the fore-limbs had to be *cephalized*, that is, taken from the office of locomotion and given over to the service of the mind, so the mouth too had to be, in a measure, cephalized. A quadruped could only have passed into a biped through a stage of arboreal life. Thus the first step toward cephalizing the fore-limbs and making them *arms* was the acquistion of arboreal habits. Emancipating the fore-limbs from the office of locomotion determined the leading features of the body, and made it human. Emancipating the mouth from the mere office of eating, and making it express the mind in speech and song, determined the features of the face and made it "the human face divine." By his *arm* man holds dominion over the world. Through his *mouth*, in a large sense, he is, and was created.

The dentist, by thrusting his forceps into our mouths, is fast making us over into a new race, a race which one day may take the name already coined for it, the name of "Lantern-jawed."

It would seem that our national malady is bad teeth. Examinations were made in the primary schools of a Boston suburb which showed that out of three hundred boys and girls under the age of twelve only fifteen had perfectly sound teeth! The dentist begins on little mouths, and his work will end in making a race with little mouths. If any one will examine a museum of skeleton heads collected years ago, as that of Dr. Morton in Philadelphia, he will see that, generally, where teeth have been extracted the pits have not closed up. The explanation is that, in general, the teeth have been lost later in life. If now he were to examine such a museum as might be made out of living heads he would find that, generally, more teeth had been lost, and more of the alveola, or pits, had been closed up. When a jaw loses a tooth and is so young and vital as to mend the chasm, finding itself too large for its uses it con-

Fig. 77.

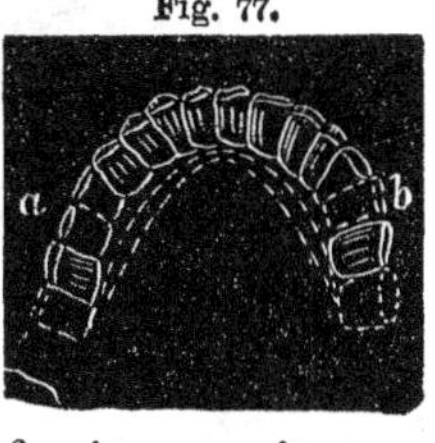

tracts. Fig. 77 represents a full, round, old-fashioned jaw, large enough to hold its complement of sixteen teeth. Suppose that early in life the teeth, *a* and *b*, have been extracted. The jaw, being too large for its needs, would tend to contract and lose its roundness of outline. It would assume the form indicated by the dotted line. It would become more triangular and pointed. It would assume the form meant to be described by that ill-chosen word, lantern-jaw.

This contraction of the jaw becomes hereditary. It is the concurrent testimony of American dentists that our teeth when sound and in their places, crowd each other for room. The jaw is too small. If nature meant it to carry sixteen teeth she has not the intelligence to measure space against number. And

Fig. 78.

here the dentist comes to help her again. He pulls out one or more of her sound teeth to give room to the others — and to let her contract the jaw a little more. The face we present in Fig. 78 is not intended as a caricature of the forth-coming "Lantern-jawed race" of America.

We return to the radical of our race-tree. The branches which symbolize the Jew, the Mongol, and the Aryan, cannot unite on this side of history, but they may unite on this side of humanity. The trunks from which so many boughs have sprung, the two trunks which symbolize the straight-haired and woolly-haired races *could not unite on this side of humanity.* The race which would generalize these earliest and most divergent trunks was pre-human. The animal was in great part eliminated from the body, since the forelimbs were already cephalized, although the hind-limbs had not developed calves. The face was still animal since the mouth was not "cephalized," but was devoted chiefly to its primary function and was more the servant of the stomach than the brain.

From this radical there sprang — through what cause we do not know — a race with woolly hair, and another with straight hair. The woolly-haired race diverged — through what cause we do not know — into a *fleece-haired* race and a *tuft-haired* race. The tuft-haired people becoming widely dispersed, separated, in time, into three sub-races — the Papuan, the Hottentot, and the Bushman. The fleece-haired people, spreading over a large portion of Africa, separated into two sub-races, the Kaffre and the true Negro. The straight-haired people spreading chiefly over the north hemisphere, but finally over all the world, separated into many races and sub-races, and in some of its branches rose into civilization, and is to rise into a noble, a divine humanity.

To de-animalize man and to develop civilizations

has been the greatest achievement in the history of our planet.

After man had become man, the first factor in the evolution of races and civilization was *dispersion*. If man were not so old here would be a difficulty. How could we account for his dispersion from island to island, from continent to continent, over wide stretches of ocean, in times which antedate civilization, antedate the crudest attempts at navigation? From Hindustan he found his way into America while yet he was a miserable savage unable to traverse an expanse of ocean. From Africa to New Guinea, or from New Guinea to Africa, or from some land which lies now on the bed of the Pacific and Indian oceans, to Africa and New Guinea and New Caledonia and the Philippine Islands, and even to Tasmania, he found his way as a tuft-haired, dark-skinned savage, no more able to build a ship than to write the Principium. But man is older than the Indian Ocean, older than Behring's Straits. His antiquity is so great that to account for his dispersion we may imagine such a collocation of land and sea as the exigencies require. To account for the dispersion of the Papuans we find Prof. Owen hypothecating a pre-historic race of *bipeds*, (the word is his,) existing age after age on a continent which in the course of gradual, non-cataclysmal, geologic change, has been broken up into islands.

Dispersed over wide areas of the globe, man is subject to different conditions and is acted upon by different forces. He begins to undergo different modifications and to break up into different races.

What he will *become* will be determined chiefly by where he *is*. To trace him now as he—

"Throve and branched from clime to clime,"

(or did n't thrive,) we are to remember first and always that he is *poor*.

MAN IS POOR. Let him toil and moil every day of his life and every toiling hour of every day, still he is *poor*. If some higher intelligence could look in on the toiling millions of Massachusetts, grinding and spinning and weaving in the mills, boring and blasting and pounding in the quarries, in the sweat of their brows torturing the unwilling earth for its grain, in sweat and grime torturing metals in the furnace, a cloud of sadness might pass over his brow as he thought that for all this toiling and moiling, after clothing their nakedness and feeding their hunger, there was left just *three cents a day to each toiler*.* Men here and there may be rich, but man is poor, for nature is so reluctant with her gifts. Gifts? She has no gifts for man except ground for his feet, water for his lips, and air for his lungs. All else he must wrest from her with the strong arm or the quick brain.

With capital of brain-work and arm-work accumulated through hundreds of milleniums, man is still poor. How poor and wretched he must have been in the beginning! An Ethnologist who has lately cruised among the South Sea Islands, and studied their people, moving among them and winning their confidence, tells us of their poverty, their misery, and their discontent. The poor wretches would beseech him to

* This is the average of surplus earnings in Massachusets.

take them away in his ship; to take them *anywhere*, if only he would take them from their own miserable homes.

Well, we are to imagine man at the start a poor, unclad, hungry, houseless vagabond. He is a troglodyte and houses in caves. He is hungry and eats whatever he can get, but he has no means yet of capturing swift-footed or ferocious game. His diet would be fruit — such poor fruit as nature herself could make — nuts, roots, shell-fish, and fish. Following this stage was the hunter stage.

Now while man is a hunter, he ought, like Nimrod, to be a *mighty* hunter. His limbs should be matched against limbs as fleet as his own, his senses against senses as quick, his arm against a paw as strong. This is the time for toning up the body and quickening the senses. When the Greeks fabulated a race of men that made war on cranes, they were wise enough to make them pigmies and poltroons. If men in an early stage of savage life were living on a land which nourished no enemy more dangerous than cranes, they would develop no vigor of body and no courage of soul. Islands sustain no large, or strong, or fleet, or ferocious animals, and island men, in general, are weak and puny. As the development of stupid, sluggish, unwieldy Megatheriums and Glyptodons in South America implies a poor concurrent development of the higher Carnivores, so the poor development of all the animal types in the Pacific Islands and in Australia implies a poor development of the animal *man*. The most puny, flimsy, impotent savage on the globe is the Australian, and the poorest fauna on the globe is that of Australia. Like nature, like man. When man spread over that land of arrested development, *his* development was arrested.

To the hunter succeeds the shepherd stage. In hunting the animal, man observes its habits. Some animals, he sees, are solitary and others gregarious. Their instinct is to follow and obey a leader. That instinct is easily diverted toward a leader not of their own species. Hunters learned not to slay indiscriminately, but to spare such animals as they could teach to own their kingship, to feed them with their milk, or clothe them with their wool, or assist them in their chase of other animals. When we remember that even ants have got on so far as to have domestic animals, it would seem no great achievement that man should have acquired dominion over animals and become a shepherd. But a great achievement it was. It was the first step toward civilization. For if the hunter stage is that in which the body gets strength and the senses quickness, the shepherd stage is one in which the *mind* is enlarged and quickened. Man becomes now what, as a root and nut and mollusk-eater, or as a hunter, he could not have been. He becomes a man of leisure. Lazily watching his flocks, he takes note of things. He considers the stars, whence they came. He considers the lilies of the field, how they grow. He becomes *circumspective*, looking around, and *introspective*, looking within. If he would become a tiller of the soil and a founder of States he must first be a shepherd of flocks. Humboldt, while surveying the herds of vicugnas roaming over the vast pasturage of South America, and ready to be shepherded, felt as one who might be presiding at the birth of States.

Look, now, over the world. Islands have but few animals that man could domesticate. The Sandwich Islands have a wild hog which might have been tamed, but a tamed hog is poor stuff on which to build a civilization. Tasmania had her "Devil,"* and Australia her Wombats and Kangaroos, but these great islands, having no animal to serve man by *killing* him now and then, had none to serve him by bearing his burdens or giving him milk or wool. In North America there had been animals, horse-like, camel-like, and pig-like, which man might have domesticated, but they came too soon or he too late. If man had been in America during the Miocene age he might have straddled and harnessed the Anchitherium as we ride and harness its descendant, the Horse.

Man must have reached Australia, Tasmania, the Pacific Islands, and North America before he rose to the rank of a shepherd, else in his migrations he would have been followed by his herds. His herds he got in Western Asia. He has domesticated about thirty species of Mammals, and the original home of nearly all of these we find to have been Western Asia. Here, then, man having been "a mighty hunter before the Lord," and having acquired strength of limb and quickness of sense, became a shepherd and a thinker. Here he first knew himself and named himself, and the name which stands to this day in the dead speech of the dead race of Aryans — the first name known to have been given to man by himself — is "THE THINKER."

In subduing the animal, man only in part subdues

* The Dasyurus ursinus.

himself. He must subdue the earth. He must become a farmer.

We look again over the world. On some of the Pacific Islands we find the taro, the yam, the sweet potato, and the bread palm. To the Sandwich Islander a little agriculture was possible, and that little had to be achieved without animal help. In South Africa there was not a weed, or grass, or berry, or fruit, serviceable to man. In Australia, there was not a weed which man could raise into a plant, nor a grass he could raise into a grain, nor a shrub on which he could create a fruit. In America there was a shrub potential of apples, a tuber potential of the potato, and a grass potential of corn. Agriculture was possible, and some of the tribes had advanced, without help from animals, so far as to cultivate patches of garden.

Man must have spread over Australia, Tasmania, the Pacific Islands and America, before he had begun to till the earth, else he would have carried with him his grains and his fruits. Most of his grains and fruits he got in Asia. Here, then, man, having been a mighty hunter and having acquired strength of limb and quickness of sense, and having become a shepherd and named himself a *thinker*, here he became a tiller of the soil and a founder of States. Here nature consented and the Aryan passed up into civilization.

Almost at the Aryan's antipodes, on a cold shoulder of the Peruvian Andes, nature favored, but in less degree, and here the American Mongol rose toward civilization.

In South America the animal kingdom is dwarfed,

and Man, as a hunter coping with nothing more dangerous than the Jaguar, did not acquire strength of limb or ferocity of will. But South America had one animal which she received as she received man, from the north. Very long ago a race of camel-like ruminants originated along the slopes of the new-risen mountains of Wyoming and Colorado. From that father-land the Procamelus seems to have wandered into Asia, where, developing into a Camel, it became an important factor in civilizing the Asiatic races, and into South America, where developing into a Llama, it became the chief factor in civilizing the Peruvian. It is small and gentle. The Peruvian laid his hand on it and it became his horse, his camel, his sheep, and his cow. Certain grasses and tubers invited, and the Peruvian shepherd became a tiller of the soil and the founder of an empire. He became an engineer of great roads and canals and a builder of great temples.

The Peruvians were murdered, the whole race murdered by Spain. Peeled they were, and scattered. From homes, from temples, from way-sides, from glades of the forest, from glens of the mountain, from lagoons where no foot had pressed save the Crane's and Flamingo's, there went up to heaven a mingled wail of babes and mothers and men, shot, stabbed, burned, murdered, for greed of gold and love of Rome. There is a legend that in a battle before the gates of Rome, the slain Huns and Romans leaped from their bodies and renewed the battle, ghost to ghost, serried ranks of ghosts on tented fields of air, under filmy banners and with filmy spears and javelins dramatizing in immortal fury the mortal combat below! If the attributes of the soul survive as the soul itself, if wrong stings till retribution heals, imagination can picture the Peruvian pavilioned on other fields, bannered and sworded for other combat, and uplifting to the Justice which slumbers long, his imprecations against the old world bigotry that slew the new world civilivation.

In Western Asia and in Peru nature permitted and Man advanced. In Australia and the Pacific Islands his way was barred at the first stage for want of animals to hunt and fight and thus to develop thews and sinews. In North America, and again in the Islands and in Australia, his way was barred at the second stage for want of animals to tame, and by giving him the leisure of a shepherd to develop in him introspection. In South Africa, in Tasmania, and again in Australia, his way was barred at the third stage for want of plants to make him a tiller of the soil. In Equatorial Africa he was wilted by heat. In Greenland he was pinched by cold. In Arabia his path was through seas of burning sand; in Siberia, through deserts of snow. Everywhere the poor step-child of nature, it was only where nature spread under his feet neither frozen snows, nor burning sands, nor asp, nor adder, nor deadly malaria, nor withheld her tribute of sociable herds, nor refused to grow her serviceable plants, that man could lift himself into a tiller of the soil and a founder of States.

We say a founder of States, for man's social evolution keeps pace with his mental, and the plow must precede the State.

While hovering on the confines of brutedom man has no other social organization than that of the animal. It is *class*, and is founded on sex. Advancing a little he organizes the tribe, founded on kinship. In its early stages the tribe is simply a number of consanguinea. Advancing still more he organizes the State, founded on the idea of territory. All transitional periods are critical, and the Australian, arrested so near the confines of animalism, having retained something of the animal in his face, retained the animal in his mind. He did not make the transition complete between the class and the tribe. The Australian is like a fern whose growth is arrested before it has sloughed from its root the liverwort thallus from which it sprang. Sticking to the root of the Australian tribe is the old thallus of the class. One-fourth of all the native women are wives at sight of one-fourth of all the men. Here is promiscuity, the promiscuity of classdom, under a single restriction. "Free-lovers" are right in picturing a social state in which the passions — "affections" they call them — are exceedingly free, right if they change the tense. Their ideal is the *first* term, not the last, in the evolution of societies.

Another critical period is the transition from tribe to State. No Asiatic people ever made the transition complete. The Jew talked about the Gentile world while he was the very Gentile of Gentiles. He never outgrew the *gens*, or tribe. It took all the force of the Greek intellect to break up the *gens*, and all the subtleties of Roman law to hold what the Greek had won.

The stages we have sketched are not separated from each other by sharp lines. There are no sharp lines, either in nature or history. But to attain civilization and make all the transitions complete, a race must pass through each of these stages.

What is the trouble with Ah Sin? Ah Sin is like a schoolboy who has passed too abruptly from the infant school into the high school. Ah Sin needed a longer schooling in shepherd life. What can you do for him? Even what he himself declares that he can do for his brother, the untamed and untameable mountain Ah Sin, *nothing at all.* The wild and savage Chinaman is a case of development arrested at an early stage. The tame Chinaman is a case of development arrested at a later stage. For arrested development there is no cure, unless it be a surgery that cuts into the bone. Here is a man with a cleft palate. What can you do for him? The time is when we all have the cleft palate. The time comes when nature, in her miracle of creation, weaves up the cleft. But if the appointed time passes and she has not closed the cleft, if in the play of her shuttle a thread is dropped or broken, she cannot go back on her path and pick up the dropped thread in the life-garment she is weaving. No more can she go back and pick up a dropped thread in the warp and woof of civilization.

One thing may be done for the man who retains the cleft palate of the embryo. The surgeon may stimulate the growth of the parts by cutting and hacking and gashing the bone. And one thing might be done for such an arrest of development as that of China. China might be stimulated to growth by the surgery of a great, murderous, decimating war. "Happy," said Montesquieu, "is that people whose annals are written in the sand." Happy, it may be, after a sort, but sluggish and stupid. For China you can do nothing — unless you do what the physician does for the cleft palate, stimulate the arrested course of development by surgery.

What can you do for the American Mongol? for the Pacific Islander? for the Hottentot? for the Papuan? for the Negro? What can we do for these lower phases of arrested development?

From the shores of the Mediterranean civilized man has spread over all Europe. He has pushed his way over both Americas. He has penetrated the dismal wastes of Tasmania and Australia. His colonies are a fringe of light on the dark borders of Africa. Wherever he has gone he has carried with him the whole serviceable world. To the savage of every tribe he has taken the animal already tamed, the weed already raised into a plant, the metal already mined and wrought into a tool. Why, then, should not the savage throw off the animal mind and pass right up into civilization? Simply because it would accord neither with law nor gospel. A man must work out his own salvation, and so must a people. You cannot make a man of your child by taking on yourself the discipline through which he should pass. The law is the same for a people. For itself a people must tame the animal, reclaim the weed, and mine the metal.

For the savage we can do nothing. His opportunity is past. He is chronic. Time was when the leopard might have changed his spots and the Ethiopian the color of his skin, but not now. We read that a time comes in the life of a man when his spiritual forces harden and he cannot turn, but must live on forever under the sentence, "he that is unjust, let him be unjust still, and he that is filthy, let him be filthy still." So, too, rigidity may overtake a race. Its summer may pass and autumn may find it leafless, fruitless, sapless. If it lives on it must live under the doom, "he that is a Modoc, let him be a Modoc still; he that is an Ethiopian, let him be an Ethiopian still, and he that is an Australian, let him be an Australian still."

But Mongol and Ethiopian and Australian will not live on, sapless branches on the family tree. Only the topmost branches, and a single branch low down toward the root, are alive and full of sap. We look at Asia and see, everywhere, the white races pushing against the olive and the brown. In China we see an empire of imbecilities, caught in the net of their own conservatism, and doomed to die because they will not grow. If we turn to the islands of the Pacific we behold humanity everywhere smitten with death. In North America we see the Indian fading like a snowbank under the suns of June. In South America we find him melting like a glacier in the breath of August.

Careful studies on the anatomy of the races have shown that among the uncivilized, and those of arrested civilization, there are some which have undergone degeneration, others which have reached the limit of culture possible to them, and a single one, that of the Kaffre, which has capacities for improvement. For only one of the races which are still under the thrall of nature and superstition science permits us to hope.

To the disenthralled races she brings a gospel full of hope and cheer. Man took this world when tenanted only by wild weeds and wild beasts, and himself, a wild man. Thorns and thistles, claws and fangs, asps and adders, typhoons and simoons and siroccos, war of beast with beast, and wind with wave—that was the world in times pre-human. No justice, no mercy, no pity was here, but war of each against each, and the elements against all. Through this very war there emerged a being destined to tame the fury of beasts, to tame even the fury of the elements, to bring peace and recreate the world. At first he was in the list of battle, the level antagonist of pard and panther. Warring against the beast, he learned to subdue it. Taming the beast, he found that he was taming the beast in himself. Tilling the soil, he found that he was tilling another soil in himself. Gaining dominion over nature, he was gaining dominion over the passions of his own nature. At last, through friendly help of herds and plants, and elements tempered more kindly to his needs, he was disenthralled, and from being a serf he became a creator. Races, arrested in their growth and held under thrall, are as exhalations—

> "Which, chilled in soaring from the plain,
> Darken to fogs and sink again."

Races, not arrested in their growth, and becoming emancipated from the thrall, are as the exhalations which—

> . . . . "Triumphant, spread
> Their wings above the mountain's head,
> Become enthroned in upper air,
> And turn to sun-bright glories there."

The ascended vapor creates all the beauties of cloud and the burnished glories of sunset. The ascending races of men will reach the cerulean heights and create a heaven of earth. Already our hands are laid mightily on the earth and the elements. Our railways are cut through mountains. Our shipways are dug between oceans. Our command is on the lightning. At our bidding the floral world has put on richer hues and sweeter fragrance. At our bidding is the fragrance of flowers where flowers are not. We create the odor of every flower that blooms save only the jasmine. We make the luscious apple from the bitter crab, and we are learning to make such compounds as the apple directly from the elements. Already we have created half the organic compounds, and when we shall learn to create them all, no victim will bleed to give us meat.

We are re-creating ourselves. We have worked the downward slant out of the bodily eye, and we are working it out of the spiritual eye. We are still in the making. Behind us, unnumbered ages of preparation, within us unspeakable potencies, before us—

"The highest mounted mind
Still sees the sacred morning spread,
The silent summits overhead."

# INDEX.

A.



B.

G.

H.



I.



K.

## L.



## M.

## R.

**PINKERTON.—Claude Melnotte as a Detective, and Other Stories.** By ALLAN PINKERTON. Each narration is thrilling. The Pinkerton series is peculiarly valuable. His books are unique. They furnish information never before given to the public. Pinkerton is the foremost detective in the country. There is a fascination in his method of narration, as well as in the subject matter of his stories. Price, $1.50.

**PINKERTON.— The Detective and the Somnambulist.** By ALLAN PINKERTON. Sumptuously illustrated with engravings by Earlie, from drawings by Mons. Felix Regamy, the great French artist and delineator. Bound in richly decorated cloth, with black and gold ornamentation. This volume embraces the thrilling story entitled The Murderer and the Fortune Teller, and is uniform with the author's previous works, "The Expressman and the Detective," and "Claude Melnotte as a Detective"— volumes which maintain their great popularity and their position as the best selling books ever issued from the American press. Price, $1 50.

**SHUMAN.— The Loves of a Lawyer; His Quandary, and How it Came Out.** By ANDREW SHUMAN, Editor Chicago Evening Journal. A volume dainty and attractive, well worthy a place in any library, as well for its elegant appearance, as for its genial and hearty matter. Elaborately bound in muslin, of different colors, with black and gold ornamentation, vignette illustrations, and vermillion edges. 75 cents.

A sprightly story, and is fragrant with country life. It is quiet, and yet the reader becomes as absorbed in the question "which do I love" as in the most exciting novel. The Loves of a Lawyer is the most popular book of the year. It will be found on every book-seller's counter, and the gentlemanly newsboys on the trains will take pleasure in showing so beautiful a little volume.

**GRIMES.—The Mysteries of the Head and the Heart Explained.** By J. STANLEY GRIMES, the Popular Lecturer. Illustrated with upwards of one hundred original drawings, engraved expressly for this work. Full of new ideas; an improved system of phrenology; a new theory of the emotions, and an explanation of the mysteries of Mesmerism, trance, mind-reading, and the spirit delusion. You hear and read about Spiritualism, Mesmerism, Ghost-seeing, Clairvoyance, and wonder whether there is any reality in these things. You naturally ask: Can I see and converse with friends in the Spirit Land? Is there any truth in Spiritualism? How can I induce the trance state? These and many similar questions are all answered carefully, correctly, and clearly, in Prof. Grimes' book. This volume contains important information to be found nowhere else. Price, $2.

**McCLELLAND.—Civil Malpractice.** By M. A. McCLELLAND, M.D., of Knoxville, Ill. Complete in one volume. Mailed free on receipt of price, $2.

"Malpractice may conveniently be divided into two kinds, viz.:—1. Civil Malpractice, in which patients bring suits for damages which they have, or think they have, sustained. 2. Criminal Malpractice, in which the People or State is made the plaintiff. The present work is limited to the consideration of the first division of the subject."

CONTENTS.—*Civil Malpractice.* Definition of Term—Allegation of Plaintiff—Definition of Ordinary Skill—Quacks, definition of—Adjudicated Cases—The Physician must Adhere to his Adopted Method—Medical Books in Evidence—Necessity for Making Visits—No Departure from Course of Treatment Adopted—Initial Bandage—Contributory Negligence—Negligence and Skill from a Medical Stand-point—Crepitus, Difficulty of Diagnosis—Dislocations—Skill in Treatment—"Setting" of Fractures, Time for Reduction—Means to Prevent Shortening—Impacted Fracture—Rules for the Government of the Surgeon in Treatment of Fractures—Amputations—Bibliography.

**POWERS & JOHNSON.—The Complete Accountant.** Designed for the use of schools and private study. By M. R. POWERS, Principal of the Commercial Department, Metropolitan Business College, and M. R. JOHNSON, author of "Accountant's Guide." Price, $2.50.

*** The most exhaustive and thorough treatise on Practical Bookkeeping ever published.

**SWING.—David Swing's Sermons.** Bound in beautiful English cloth. Price, $1.50.

Prof. Swing's best Sermons. "He fascinates the careless and the unconverted, and wins them over to Christianity." A cheap and beautiful volume. An account of the Trial, and "Declaration" presented to the Presbytery by Prof. Swing, in explanation of his theological opinions.

DAVID SWING'S CHURCH.—THE CENTRAL CHURCH.—*Its Creed.*—We, whose names are signed to this paper, earnestly desiring to promote our own spiritual welfare, and to take some part in the great work of helping others to lead the Christian life, living in a large city, where the moral work to be done is so great, do form ourselves into a Christian society, to be known as the Central Church of Chicago. We would found our church upon the great doctrines of the New Testament. We believe in the divine character and mission of Christ; that He is the Savior which man in his sinfulness and darkness needs; that all those believing and following this Christ are entitled to the name of Christians. Furthermore, as at the Holy Communion many leading evangelical churches cordially invite to the Supper all who love the Lord in sincerity and truth, so we, feeling that no service of the sanctuary is holier than its communion, would invite into full membership all who make this Savior their way, truth, and life.

**ARTHUIS.—Static Electricity.** A new book on Nervous and Rheumatic Affections treated by Static Electricity. By Dr. ARTHUIS, a noted Practitioner of Paris, France. Translated by J. H. ETHERIDGE, M.D., Professor of General Therapeutics in Rush Medical College, Chicago. One volume, 12mo., illustrated. Price, $2.

CONTENTS.—History of Medical Electricity—Inferiority of Dynamic Electricity, its Dangers—Static Electricity; Operative Procedures—Electric Machine—Insulator—Excitators—Electric Bath—Absorption of Electricity by the Human Body—Electric Currents—Sparks and Electric Frictions—Electric Douches and Pulverizations—Natural Electric Currents—Frictions, Shampooing—Examination of the Patient—On Transport of Medicines by Static Electricity—Medicines—Experiments of Dr. Burq—Clinical Observations—Epilepsy—Paralysis with Aphasia—Progressive Locomotor Ataxia—Rheumatism—Muscular Contractions—Rheumatism, General Innervation, Moral Prostration—Hysteria—Chorea, or St. Vitus' Dance—Neuralgia—Facial Neuralgia—Intercostal Neuralgia—Sciatica—Gastralgia, Nervous Vomitings—Asthma—Pulmonary Emphysema—Deafness—Amaurosis—Action of Static Electricity on Menstruation, Dysmenorhœa—Chronic Diarrhœa—Incontinence of Urine—Paralysis of the Bladder—Tonic and Recuperating Action of Static Electricity upon Enfeebled Subjects and Old People—Hypochondria—Hectic Fever—Pulmonary Phthisis.

www.ingramcontent.com/pod-product-compliance
Lightning Source LLC
LaVergne TN
LVHW010136110826
845151LV00002B/361
*9781425539603*